AS THE MINUTES TICK AWAY...
HELL STAKES A CLAIM
IN MARTINIQUE...

This is the minute-by-minute, hour-by-hour, day-by-day story of an incredible happening...

Friday, May 2
The first victim, the first warning

Saturday, May 3
Tumbled from bed, the world in flood

Sunday, May 4
A mountain on the march, the sky darkens

Monday, May 5
The great wave

Tuesday, May 6
Requiem for the living

Wednesday, May 7
The end of the affair

Thursday, May 8
THIS DATE SHOULD BE WRITTEN IN BLOOD

PAUL NEWMAN
JACQUELINE BISSET • WILLIAM HOLDEN
in
IRWIN ALLEN's
production of

WHEN TIME RAN OUT...

EDWARD ALBERT • RED BUTTONS
BARBARA CARRERA • VALENTINA CORTESA
VERONICA HAMEL • ALEX KARRAS
BURGESS MEREDITH
and
ERNEST BORGNINE and JAMES FRANCISCUS
as Bob Spangler

AN INTERNATIONAL CINEMA CORPORATION
PRESENTATION

Based on the novel
THE DAY THE WORLD ENDED
by GORDON THOMAS and MAX MORGAN WITTS

Screenplay by
CARL FOREMAN and STIRLING SILLIPHANT

Produced by
IRWIN ALLEN

Directed by
JAMES GOLDSTONE

Distributed by Warner Bros.
A Warner Communications Company
© 1980 WARNER BROS. INC.

When Time Ran Out...

Gordon Thomas
and
Max Morgan Witts

Originally published as
The Day the World Ended

*This low-priced Bantam Book
has been completely reset in a type face
designed for easy reading, and was printed
from new plates. It contains the complete
text of the original hard-cover edition.*
NOT ONE WORD HAS BEEN OMITTED.

WHEN TIME RAN OUT ...
*A Bantam Book/published by arrangement with
Stein & Day Publishers and Souvenir Press, Ltd.*

PRINTING HISTORY
Originally published as "The Day the World Ended"
Stein & Day edition published January 1969
Literary Guild edition December 1968
Serialized in the DETROIT NEWS *and the*
NATIONAL ENQUIRER *1969*
Condensed in READER'S DIGEST *April 1972*
Bantam edition/April 1980

All rights reserved.
Copyright © 1969 by Gordon Thomas and Max Morgan Witts.
Cover art © 1980 Warner Bros.
*This book may not be reproduced in whole or in part, by
mimeograph or any other means, without permission.
For information address: Stein & Day Publishers,
Scarborough House, Briarcliff Manor, New York 10510.*

ISBN 0-553-14099-X

Published simultaneously in the United States and Canada

Bantam Books are published by Bantam Books, Inc. Its trademark, consisting of the words "Bantam Books" and the portrayal of a bantam, is Registered in U.S. Patent and Trademark Office and in other countries. Marca Registrada. Bantam Books, Inc., 666 Fifth Avenue, New York, New York 10019.

PRINTED IN THE UNITED STATES OF AMERICA

St. Pierre, Martinique, May 8, 1902.

"This date should be written in blood. . . ."
—M. Gabriel Parel,
Vicar-General of Martinique

Acknowledgments

Archives Nationales, Paris.
American Embassy, London.
British Broadcasting Corporation.
British Museum, London.
British Geological Survey.
Catholic Missionary Society.
Cable and Wireless Ltd., London.
Foreign Office, London.
French Embassy, London.
French Institut du Royaume Uni, London.
Lloyds of London.
National Archives, Washington, D.C.
Public Records Office, London.
Royal Empire Society, London.
Tate and Lyle Refineries, Ltd., London.

Special Thanks

Madame Anca Bertrand, Fort-de-France.
M. C. F. Beauregard, Le Secrétaire Général de la Chambre de Commerce et D'Industrie de la Martinique, Fort-de-France.
Monseigneur Henri Varin de la Brunelière, Bishop of Martinique, Fort-de-France.
Madame Marie Dufrénois, Morne Rouge.
M. Joseph Durival, Conservateur du Musée Volcanologique, St. Pierre.
M. Pierre Lambertin, Le Préfet de la Martinique, Fort-de-France.
M. Pierre Milon, Fort-de-France.
M. Raphaël Petit, Le Préfet de la Haute Loire, Le Puy.
M. Eugène Pierre-Charles, Le Maire de la Ville de Saint-Pierre.
M. J. Petit Jean Roget, Le President, Société d'Histoire de la Martinique, Fort-de-France.
M. Bernard David, Fort-de-France.

Friday, May 2, 1902

Contents

PROLOGUE

FRIDAY, MAY 2, 1902
1. The First Victim — 5
2. The Anxious Bridegroom — 14
3. A Cell with a View — 26

SATURDAY, MAY 3, 1902
4. Tumbled from Bed — 39
5. Lava in the Throat — 54
6. The Whole World in Flood — 71

SUNDAY, MAY 4, 1902
7. Special Powers — 79
8. A Black Mass — 88
9. The Prison Riot — 100
10. Released to the Living — 110

MONDAY, MAY 5, 1902
11. The Pavement Slaughterhouse — 127
12. A Mountain on the March — 138
13. The Great Wave — 147
14. A Shortage of Coffins — 156

TUESDAY, MAY 6, 1902

15.	The Sympathy of the Government	165
16.	A Handful of Francs	176
17.	The Deepening Crisis	188
18.	A Fateful Reprieve	195
19.	Stars and Stripes	208

WEDNESDAY, MAY 7, 1902

20.	Gabriel's Angel	219
21.	The End of the Affair	232

THURSDAY, MAY 8, 1902

22.	*"Allez"*	247
23.	Requiem for the Living	258
	EPILOGUE	281
	BIBLIOGRAPHY	285
	AUTHOR'S NOTE	290
	BACKGROUND TO THE BOOK	291

Prologue

THERE HAD BEEN a rumbling for two weeks. It was deep-throated and muted, coming as it did from the bowels of the earth. It had caused no panic in the town. The noise and the puffs of smoke that accompanied it were reported by the local newspaper as being of no more than passing interest. Even at the beginning of the third week, when the night sky was lit by flashes of light that resembled artillery fire seen from a great distance, nobody in the town was unduly worried. The most positive reaction was from the man who sat down and wrote a letter to the London *Times.* He noted that the rumbling, the smoke and the ash that it generated, and the light flashes "all seemed to be produced so normally that even those who are inclined to look on the dark side seem to have nothing to fear."

In that month of April there was indeed no one in the town who appeared frightened.

The first day of May dawned to produce a lull in the rumbling and its attendant side effects. The townspeople smiled at each other; on the whole they were convinced that they had been right in not fearing the happenings of the previous twenty-three days.

Then late in the afternoon they saw that the mighty cone from which the puffs of smoke had been emitted was now presenting a new phenomenon. Vapor trails careened into the sky, criss-crossing each other to form intricate patterns. They produced interest, speculation, even excitement among the younger population. But again, there was no recorded case of fear.

The vapor trails were still visible that night when the town went to sleep.

It was in the early hours of May 2, 1902, that they ceased. Shortly afterward the fallout came, swiftly and silently, to St. Pierre. The final agony of a town that some believed deserved to die had begun.

FRIDAY

May 2, 1902

CHAPTER ONE

The First Victim

HE WAS ALWAYS PUNCTUAL. Each morning when Fernand Clerc rose from his brass bedstead—which like the cotton nightshirt he wore had been imported from France—and pushed open the stout shutters of his bedroom, the inhabitants of St. Pierre knew it was six o'clock. As he walked out on the balcony which ran around the house, the distant toll of the bell of the Cathedral of Saint Pierre carried across the awakening city and up to the hillside where he stood.

A remarkable sight greeted him. The entire town, the sea beyond, and the thick hinterland appeared to be coated with a heavy fall of hoarfrost. Yet the sun was hot on his skin. Nearer at hand, inside the estate wall, on the branches of the lime trees, Clerc noticed that the frost appeared to be like crude cement powder; below him, in the paved courtyard, it lay thick and even. The powder completely hid the waxed floorboards of the balcony. Where it had been caught by the wind that blew all year over and around the island, the powder had drifted to form conical shapes that were not unlike volcanoes.

Clerc stooped down and pinched the tip of one of the mounds between finger and thumb, rose to his feet, and carefully sniffed the powder. It had a bitter aroma about it that he instantly recognized.

He had last sniffed it a month before. Then, he had gone with his family to the L'Étang des Palmistes, a lake in the crater basin of Mount Pelée, the island's largest volcano. They had gone together, on a Sunday afternoon, for a picnic by the lake. As they had sat there, inside the rim basin of the volcano, watching the clear light casting giant shadows on the pumice rocks, Clerc had noticed a wisp of smoke rising from a corner of the lake.

With his two children, he had walked across the gritty pumice soil to investigate. As they approached, a breeze sent the wisp curling toward them. The smoke had made their eyes smart, and stung the backs of their throats. Then, as they stood there, coughing and spitting, the water had bubbled, and a black jet had fountained into the air. It fell in a lazy arc across the branches of a tree overhanging the lake. They looked, smarting and rasping temporarily forgotten, at the now seared and withered branches. Without waiting to investigate further, Clerc had hurried his family away from the crater.

Three weeks ago, alone this time, he had returned to the volcano. From a distance, Mount Pelée—the "bald mountain"—was the classical volcanic shape: a circular cone culminating in a single summit peak that sloped on three sides down to the sea. Closer, the bold strokes of Nature's erosive carving were revealed in numerous canyons and gorges.

To reach the crest Clerc had to pick his way over piles of pumice. The top resembled a truncated loaf surrounded by a circle of pointed hills. The highest, Morne de la Croix, stood about two hundred feet above the lake. Clerc picked his way up its slope to gain the maximum vantage point. For a man approaching middle age, he showed little sign of his exertions. He paused to gain his breath, and then looked down into the crater.

For a long time he stood there, unable to believe his eyes. The placid surface of the lake was now a black mixture of bituminous appearance, bubbling and boiling, rising and puffing. From time to time jets of white vapor and scalding water escaped, and then fell back brusquely into the cauldron.

Mount Pelée, asleep for fifty-one years, was almost fully awake. He had hurried back to the estate, bathed to remove the smell of sulphur from his body, and then sat down in his study to write a report on what he had seen. He had sealed the report in an envelope and addressed it to the island's Governor, Louis Marius Mouttet. He had summoned a houseboy and ordered him to go on horseback to Fort-de-France, fifteen miles down the coast, deliver the report to the Governor's office, and await a reply. When it came, it turned out to be no more than a polite formal acknowledgement with the rider that if the situation worsened, the Governor should be informed. Clerc had not been surprised by this lack of reaction on the part of the Governor; for some time their relationship had been strained.

Now, standing on the balcony three weeks after the Governor's reply, and sniffing the sulphur ash that during the night had been ejected by Pelée's crater, Clerc realized that the situation had definitely worsened.

He turned and looked away up the steep slopes to where Pelée's cone brooded in the still morning. Nothing disturbed the air around the crater. But the slopes were shrouded in ash that at a distance of two miles looked like icing on some giant Christmas tree decoration.

Clerc shook his head. Then he continued with the ritual he had followed every morning since he had moved his family from the town to this hillside home a mile outside St. Pierre. For the past ten years, every morning including Sundays, he had arisen before his family and walked on to this balcony to look down on the town whose foundations generations of Clercs had helped to strengthen. A Clerc had been among the first shipload of French who had settled the island of Martinique in 1635.

The island is part of a group known as the Lesser Antilles, stretching—"like piers of a bridge across the Seine," wrote an early settler—across the entrance to the Caribbean Sea. In the shape of an arc bowed out toward the Atlantic Ocean, they extend for 450 miles from the Anegada Passage, near the Virgin Islands, southward almost to the coast of South America. The northern part of the group are called the Leeward Islands; the southern half are called the Windward Islands. All of them are the result of volcanic eruptions forcing the ocean bed up 10,000 or more feet. From prehistoric times this string of islands has stretched across the throat of the Caribbean like a chain of smoldering furnaces about to burst into flame.

Those first settlers had come ashore at a great bay that had been named Fort-de-France, a name that had a comforting, secure ring about it to those seventeenth-century men, women, and children who had sailed halfway across the world to find a new home. Fort-de-France had become the island's political capital; later, when St. Pierre had been built, the commercial center of the island had found its home in the new town.

The settlers could not have known it, but they had chosen to settle on an island that was a dormant time bomb. The fuse was buried deep inside Mount Pelée, named by the colonists after the Hawaiian goddess of volcanoes, Pele. But only twice in three hundred years had the time bomb started to tick. Once, in 1792, it had spluttered a thin scattering of ash over the crater's immediate slopes; in 1851, the volcano had erupted after a week of rumbling, throwing out a column of ash which had drifted down to coat St. Pierre in a mantle of white. The eruption had lasted only a short time. It was followed by a severe rainstorm that washed all the ash away. Few had been frightened.

Now, on this quiet morning of Friday, May 2, 1902, Clerc saw that this night's fall of sulphuric ash also appeared to have caused little concern in the town. He could make out a few early risers standing in the streets, looking toward the silent cone of Mount Pelée, towering two miles inland. From his vantage point Clerc observed no panic. Probably the townspeople had lived with the volcano for so long that they had long ceased to look upon it as a threat, he mused, and understandably so, for the town had been built on a beach at the base of a great natural amphitheater of rocks which rose to end at the summit of Mount Pelée.

St. Pierre had become the island's social as well as commercial center. It was known, as far away as Puerto Rico in the north and Trinidad in the south, as the Paris of the West Indies. It strove hard to live up to its reputation, importing cabaret acts from Paris, fleecing the sailors in the waterfront bordellos, and living at a pace that staid Fort-de-France thought positively indecent.

Clerc himself, a devout family man, had little time for

The First Victim

the night life of the town. One of the reasons he had moved out from its precincts was to escape from the late-night carousing; another reason was that he felt it more in keeping with his standing in the community to live apart from it, to view it, as he did every morning, from a distance. His vantage point enabled him to feel both dispassionate and paternal toward the city below.

This morning, even with its shroud of ash, St. Pierre offered Clerc a view he could take much pride in. From his eyrie, he had a clear view of the rum distilleries and furniture factories that were the spine of the island's economic framework, a spine whose very nerve ends he controlled. He was the largest planter on the island, a millionaire, undisputed commercial overlord of the island's population of 189,500. Next to Governor Mouttet, Clerc was the accepted social leader of the French-born colony who effectively administered the island for France. Most of them worked and lived in the streets leading off the Rue Victor Hugo, the town's main street, which ran in a gentle crescent along the sea front. Bisecting it was a series of shorter streets, leading back from the waterfront, and petering out against the slopes of the sharply rising amphitheater of rock. The cathedral, the military hospital, the town hall, and the island's two leading banks—the Bank of Martinique and the English Colonial Bank—were all clustered on the Rue Victor Hugo; but for the most part the town consisted of two- and three-story buildings, made of stone, with red tile roofs. Many of these were clustered along the side of the Roxelane, a river that ran through the town and emptied itself into the sea. Except in the rainy season it was little more than a sluggish stream; when the rains came, it became a raging torrent, swirling fish and debris from the hinterland out to sea. The bulk of the town's middle-class Creole population lived beside it. Below the Roxelane was the harbor, a natural, curving basin through which the entire commercial, financial, and social life of the island funneled itself.

That morning, Clerc counted eight ships anchored in the roadstead. Five flew the American flag. They brought from the United States the supplies essential to the island's survival. The ships' holds were filled with flour, dried vegetables, salted and smoked meats, and leaf tobacco, as well as crates containing clothes, clocks, oil lamps, and sewing machines.

The other three ships flew the British flag. Their holds were empty, waiting to be filled with the island's raw materials—sugar, rum, cocoa, and coffee—waiting to be filled with the produce of Clerc's estates that dotted the island's 380 square miles, estates that helped guarantee Martinique's economic standing in the world.

Particularly dependent on that economic standing were the bulk of St. Pierre's population of 26,011 men, women, and children who lived downtown, away from the houses of the French-born *colons*. They were the *propriets vivriers,* smallholders, vari-colored from a century and more of breeding with the sailors of the world who came to the island. Healthy exports meant a regular supply of ships calling at Martinique; ships meant customers, in one form or another, for the *propriets vivriers*. Few of them had as much as spoken to the man who lived on the hillside outside the town, but all knew their debt to his enterprise.

As the bell of the Cathedral of Saint Pierre ceased its tolling, Clerc reached for a telescope standing in a rack on the balcony. Adjusting the eyepiece, he brought into sharp focus a section of the area where the smallholders lived. Each holding appeared to be the same: a tin-roofed house surrounded by trees bearing ripening avocados, mangos, star apples, oranges, limes, and grapefruit. Studded between the trees were tufts of sugar cane, breadfruit plants, and tunnels of bamboo that sheltered root vegetables from the midday sun. Tethered around the tunnels were goats, cows, and small Creole horses. Like the rest of the town, the mulatto quarter appeared calm under its coat of ashes.

Clerc lowered his telescope and walked to a corner of the balcony. This was a new departure in his morning routine; it was one that he had started after returning from his second visit to the crater. The day after he had returned, he had secured a barometer to one of the supporting pillars of the balcony. Now, as he had done every day since then, he carefully studied the needle. It was twitching a little more than it had done the previous morning when he had inspected it. He turned again and looked up toward the mountain crater. As he did so, there was a deafening crash—"like a broadside of cannon," he was to remember—and a fresh column of ash shot into the sky.

Clerc watched it soar heavenward. Then, caught by the wind, it started to dissipate, spreading across the sky and blotting out the sun. In a minute the ash was falling silently on St. Pierre.

Mount Pelée, fifty miles square at its base, had stirred a stage further, emitting yet another warning.

Clerc turned at a noise close behind him. It was his wife, Véronique. She padded out to the balcony in her nightdress. In her hand she held a crucifix. The statue of Christ on the Cross had been carved from a piece of pumice stone taken from an upper slope of the volcano.

Together, the Clercs knelt and prayed on the balcony. Fifty-one years before, to the day, the last time Mount Pelée had awakened, Fernand Clerc's grandfather had also prayed. Then, the volcano's activity had been short-lived, and there were many on the island who had said this was due to the intervention of God in answer to the prayers of Jean Marc Clerc.

Now, his grandson and wife prayed for similar intervention.

Sitting on the doorstep of his house in the Mouillage quarter of the town on the quayside that ran along the entire water front, Léon Compère-Léandre watched the ash falling on the three British ships riding at anchor in the roadstead. The air reeked of sulphur, bringing tears to his eyes and a dry retching feeling in his throat. He pulled a handkerchief from his pocket and tied it over his nose and mouth. It offered a little relief against the acrid fallout.

Léon, aged twenty-eight, was a shoemaker. He was descended from one of the penal laborers who had originally been transported from France to carry out heavy manual work on the island. Intermarriage with Creole and Negro factions of the island's population in previous generations had now given Léon's skin a light coffee pigmentation. At a distance he could pass for a *colon*: a useful asset when it came to drinking in the water-front taverns, where tempers often ran high and the descendant of a convict was still regarded as byplay for practicing knife-fighting by many of the sailors.

Intermarriage, though, had done little to remove the basic superstitions that had been handed down to him from his father, and from his father's father—superstitions that still ruled his life. They included a belief in the Devil and that Mount Pelée was an air vent through which the Devil could breathe fumes from Hell.

Now, as the ash settled everywhere, carpeting everything in sight a dull grayish-white, Léon was convinced that Evil was afoot.

He also felt, this time instinctively, that he would not be safe in his waterfront home. He decided instead to seek shelter in his basement cobbler's shop in the town's main square, Place Bertin. Holding his handkerchief over his mouth and using the other hand to wipe the tears from his eyes, he set off for his shelter.

From sea level, St. Pierre appeared to back on to the very edge of the southern foot of Mount Pelée, rising in tiers from the open roadstead which formed its harbor. At this level, too, the streets seemed much narrower than they had appeared from Clerc's vantage point. The houses, all stone-built and stone-flagged, crowded in on each other, each with its wooden or zinc awning casting a permanent shadow over the cobbles.

Léon's sandaled feet made no sound as they scuffed over the ash. The only sound was that of running water pouring through the deep gutters of the twisting, climbing streets. This morning the water, usually clear, was stained and sluggish. The ash had already started its corrosive course of discoloration and clogging.

As he approached the Place Bertin, Léon had still seen no one; St. Pierre, it seemed to him, through either choice, force of habit, or unawareness, had decided to ignore the falling ash.

Just short of the entrance to the main square, Léon came upon a young horse tethered to a wall-ring of a house. Like all the other dwellings in the street, the shutters of this house were tightly closed beneath the traditional gable dormers. As he approached, the horse, already well coated with ash, started to snort and pant in pain.

The First Victim

Suddenly it buckled at the knees, making a dreadful, choking sound as it did so. It rolled its eyes in terror, struggled to regain its footing, found it was too far gone for the effort, and slumped down on its side.

Léon stooped beside it. One of the other superstitions that governed his life was a belief that being kind to animals would insure him a place in Heaven. He looked around for some means of giving the horse practical assistance. Nearby, at the entrance to the Place Bertin, was a water trough with a tin bucket attached to it by a slim chain. The bucket and trough were used to replenish water supplies for caravans of mules that journeyed into the hinterland. Léon tugged at the chain, exerting pressure in sharp pulls. The chain snapped, freeing the bucket.

He dipped the bucket into the trough, pushing it below the film of ash coating the surface. Then, slopping water as he ran, he returned to the fallen horse. Cradling the animal's head, he trickled water over its mouth and nose. He had wasted his strength. The horse was dead, suffocated by the falling ash.

Mount Pelée had claimed its first victim.

CHAPTER TWO

The Anxious Bridegroom

FERNAND AND VERONIQUE Clerc were not alone in seeking solace and comfort in prayer that morning. In various parts of the town, a number of people had risen early for morning Mass in the Cathedral. On Thursday, a week away, it would be the holy day of the Ascension, when people would travel from all over the island to celebrate the feast, both in the precincts of the Cathedral and later in the taverns and dining rooms of the town.

Among the early risers that morning was Mrs. Clara Prentiss, the wife of the American Consul. In her elegant home near the town's botanical gardens, she dressed carefully, mindful of her position as the wife of a senior foreign diplomat on the island. Putting on her lace-trimmed bonnet and white silk gloves, she stepped briskly out into the street.

As she did so, a distant fresh rumble came from the crater of Mount Pelée. She paused, wondering what was to follow. She had not long to wait. A column of white billowed out from the cone, and shortly afterward the ash fall descended on her part of the town. It fell thicker and faster than it had done before.

Mrs. Prentiss stood quite still, watching the ash settle around and on her. In a minute her clothes were ruined, irrevocably stained by the substance. As she stood there, she was dimly aware of several people running past. The ash deadened their footfalls; the only sound they made was a hoarse panting as the lava ash caught in their throats.

Across the street, a small bundle fell from the roof of a house. It writhed on the pavement, scuffing and slithering on the cobbles. Clara Prentiss walked across the road and saw that it was a bird. Weighed down by the ash fall, its wings had been unable to lift it from its rooftop nest, and it had plummeted to the pavement. She picked it up and started to dust the ash off its wings and body. As she did so, a man stopped beside her. "I wouldn't waste your sentiment on a bird, Madame," he said.

Before she could reply, he had taken the bird from her hands and tossed it over a nearby wall. Without another word he turned and ran off into the ash storm.

She stood there for a moment, the ash bringing tears to her eyes. Then she turned and walked rapidly back to her home, pulling off her hat and gloves as she entered. She walked briskly up to her husband's bedroom, pausing outside to look behind her. She had left a trail of ash footprints across the hall and up the stairs. She pulled open the bedroom door. Her husband Thomas was still in bed, reading the town's leading newspaper, *Les Colonies*. He put it aside as his wife entered.

Without pausing in her stride, Clara Prentiss walked to the shutters and forced them open. A flurry of ash entered the bedroom. She pulled the shutters closed and walked to the foot of the bed and faced her husband.

"It will erupt soon," she said, "it must."

For a little while there was silence in the room. Then Thomas Prentiss rose slowly from his bed, a portly, soft-muscled man, and walked to the shutters. He pushed them open and looked out at the steadily falling ash. Soon he was coughing from the tang of sulphur. Only then did he appear satisfied at what he saw. He drew the shutters closed.

"I must arrange for you and the children to leave here as soon as possible," he said.

"I am not going without you!"

Clara Prentiss was a determined woman. On May 10th, the family would celebrate their fourth anniversary on the island, though her husband had not been appointed American Consul to Martinique until April 3, 1901. To mark the occasion she was planning a quiet family dinner party for him and two of their children, Christine and Louise. Her two sons would not be able to attend. Her eldest child, James, lived in Chicago; Thomas, second in succession to the modest Prentiss fortune, lived in Batavia. Apart from the coming dinner party, Clara Prentiss believed that in her role as a diplomat's wife, it would be unthinkable for her to leave her husband's side at a time of impending crisis. It was a similar sense of position that had stopped her going on to morning Mass. In her eyes it would have been almost a mortal sin for the wife of the American Consul to appear before the Cathedral's worshipers as if she had been dragged through a cinder bath.

Suzette Lavenière, aged twenty-four, was preparing breakfast when she heard the new rumbling. It was deeper, and louder than ever before. She put down the coffeepot and hurried out of the kitchen to investigate. Outside in the villa's courtyard a number of workmen on her father's estate were grouped talking excitedly among themselves and pointing across the valley to the summit of Mount Pelée seven miles away. They fell silent at the appearance of Suzette.

Her father owned one of the island's 1,150 sugar cane plantations. The Lavenière estate was a large rectangle of land, four miles to the south of St. Pierre. The estate, rising from the floor to halfway up the side of a valley, looked across to the town and the volcano towering above it. Like many other members of the landowning ruling class of Martinique, the Lavenières ran their estate with a firm but tolerant hand. The hundred native families who worked the crop were well housed, adequately clothed, and fairly paid. They lived, at a decent distance, beside the villa Suzette occupied with her father, a widower.

That morning, Pierre Lavenière had risen early to inspect the furthest outpost of the estate, about a mile from the villa, at the bottom of the valley. He had taken his foreman and six hands with him. They had gone to investigate a report that the ash fallout had blighted the young crop and that the stream, which bordered the estate on three sides, had risen sharply in the last twenty-four hours.

Flooding was nothing new. Every year during the five months' rainy season, the stream breeched its shallow banks. But the rainy season was over, and the stream had reverted to little more than a sluggish trickle.

As he had set off, Lavenière had turned to his daughter and murmured, "It could be the mountain. The stream begins near the summit."

Then, an hour before, the summit had been peaceful. Now, as she stood in the courtyard, the very ground beneath her seemed to shake from the rumbling. A tongue of flame, orange-brown, forked out of the cone, and then was seemingly sucked back into the volcano.

Years of self-dependence had taught Suzette Lavenière to stay calm no matter what the situation.

"It is not as bad as it was earlier this morning. There is no ash falling," she said firmly.

"But what about the flames?" protested a native.

"There is no need to worry," repeated Suzette. "All the same I will go and fetch my father. In the meantime go about your work like any other day."

The words, delivered with assurance, appeared to satisfy the natives. They drifted away to their various tasks. As they did so, the rumbling and fire from the mouth of Pelée died away. A total silence appeared to settle over the countryside. Without knowing why, Suzette feared the silence more than the noise. Forgetting that she had not finished her breakfast preparations, she walked to the stable, mounted a horse, and set off to find her father. She started to sing, in much the way that, as a child, she had sung to keep up her courage in the dark.

Suzette had ridden for only a little time before she became aware of a dull roar ahead. She kicked her horse into a canter and hurried through the cane. Shortly afterward she heard the shouting, also coming from ahead, from the direction of the stream. Suddenly, through a break in the cane, she saw a sight that stunned her. The quiet, sluggish stream had vanished—in its place a sea of boiling mud was racing down the far slopes of the valley and spreading across the valley floor like a thick glue.

Firmly trapped in the filth were her father and his men. Already the mud had reached the haunches of their horses. They had been on the far side of the valley when the avalanche had poured down on them. Struggling to get clear, they had stood no chance against this tidal wave of mud. All they could do now was cry pitifully for help.

Even as she watched, Suzette knew there was nothing she could do, nothing anybody could do.

She was still screaming hysterically when an hour later workmen from the villa arrived to find out what had happened to her. By then her father and his men had sunk beneath the mud and disappeared.

Around midmorning, the inter-island steamer *Topaz* was approaching Macouba Point, a headland that jutted out from the foot of Mount Pelée. The ship had left the island of Dominica, twenty-seven miles to the north, at breakfast time, bound for Martinique and St. Lucia. She was scheduled to make three stops at Martinique: at the fishing hamlet of Le Prêcheur, at St. Pierre, and at Fort-de-France. She carried a crew of eleven and twenty-nine passengers.

On the *Topaz*'s bridge, her master, Jules Sequin, had been studying the coastline ahead for some time through a telescope. Every so often he would adjust the setting to give him a clearer view of the hinterland and the volcano overshadowing it. At last he lowered the telescope and turned to the mate standing behind him.

"Order engines slow ahead. The mountain has erupted since we were last here. It's best that we don't go too close now," said Sequin.

Born and raised on the island of Dominica, this short, broad-beamed sailor had a healthy caution for the caprices of nature. Twice in his twenty-six years at sea he had been close to death from one of her whims. Once a hurricane had wrecked the ship he sailed in; another time a waterspout had lashed his ship, badly mauling the superstructure.

Under his instructions, the *Topaz* started to turn away from the coast.

Below him, on the open deck, all the passengers were gathered, talking excitedly to each other. To them, without the benefit of a telescope, the shore seemed indistinct and blurred, the peak behind it no more than a mass of rock. But Sequin had noticed an addition to the hundreds of valleys and gorges that dropped sharply into the sea on either side of Macouba Point. It was a long tongue of fresh lava. Through his telescope he had seen steam rising from it. He had traced the lava stream back to its source at the summit of Pelée. It was this discovery which had influenced him to turn away from the coast.

What had excited the passengers was a gray film of ash floating out to sea toward the *Topaz*. Dead fish floated in the film.

Sequin sent the mate down to the deck with orders to reassure the passengers that there was nothing to fear. But at the same time he ordered the helmsman to take the *Topaz* right out to sea, and then to approach Le Prêcheur, the first port of call, in a wide arc. When he reached there, he would report in full what he had seen. Only one thing puzzled him: the dead fish. Neither the ash nor the lava roaring into the sea would have killed so many.

What he did not know was that Pelée, extending 10,000 feet to the seabed, had twitched at its base and sent a shock wave hurtling to the surface that had killed everything for several square sea miles. In the process it had also damaged the underwater telegraph cable from Martinique to Dominica.

Soon, Martinique would be totally cut off from any effective communication by cable with its neighbors.

For thirty-one years, the cables had been the island's principal method of communication with the outside world. They had been laid by the Eastern Telegraph Company under the supervision of the veteran English cable-layer Sir Charles Tilston Bright. The link from Dominica to Martinique had been opened in June, 1871; a few months later, Bright, aboard his cable ship the *Dacia*, had telegraphically joined Martinique to St. Lucia.

Now, in the next few days, these links would be broken by Pelée. The first break would come in the link between St. Pierre and Dominica within twenty-four hours; two days later, on the morning of May 5, the cable linking Fort-de-France with its southern neighbors would fail; a day later, St. Pierre would also be cut off from those neighbors. Martinique would be isolated from the world.

But before that happened several telegrams would have arrived from Paris, via Dominica, for Louis Mouttet, the Governor of the island. Those telegrams from the French Government were to have a decisive effect on the future of the people of St. Pierre; in many ways they would decide whether they would live or die.

At midday the parish priest of Le Prêcheur started to toll the church bell. It was a short, urgent summons that brought the villagers out of their homes to muster in the square in front of the church. They stood there, nearly two hundred men, women, and children, ankle deep in the warm ash.

Since dawn, when the cindery powder had started to rain down on Le Prêcheur, they had cowered in their homes, watching the fallout silting up everything in sight. It had covered the square and the two village streets. It had settled on the water-front jetty and the fishing boats moored there. It had turned the fields around the village a dull gray.

Now, with the noonday sun beating down, the whole area had taken on the appearance of a moonscape. Against this pagan background, the sight of the parish priest standing in front of their church, holding a small statue of the Virgin Mary in his hands, appeared to some villagers as incongruous.

Placing the statue at his feet, the priest urged his flock to kneel and pray for deliverance. They had hardly knelt in the ash before somebody spotted a trickle of lava nosing into the village. The impromptu service was abandoned.

The lava moved slowly, its force spent, having traveled five miles from its source at the summit of Pelée. But it still moved. And the sight of it seeping into the village, brown and gently steaming, brought the first real panic to Martinique.

The villagers, led by their priest, fled from Le Prêcheur, taking the twisting coastal track to St. Pierre, seven miles to the south.

The Anxious Bridegroom

Behind them they left in the deserted square the statue of the Virgin Mary. Later, when the lava crept into the square, it lapped against the base of the statue and finally toppled it into the mixture.

In St. Pierre the latest edition of *Les Colonies* offered reassurance to the population. Its front page announced that an unnamed "leading authority" on volcanoes had told the newspaper that the ash fallout was no more than a passing phase and that shortly the crater would become dormant again. As a footnote, the newspaper added that this coming Sunday it would organize "a grand excursion" to the summit so that readers could see for themselves the "true position."

Not one reader suspected that the paper's volcanic expert was none other than the editor Andréus Hurard, who had deliberately concocted the "authority" for political reasons. It was the first of many examples of political expediency that in the days to come would fatally overrule all other considerations in St. Pierre.

As he stood in his office late that afternoon looking out over the Place Bertin, Hurard wondered how long he could keep up what would later be regarded as a gigantic fraud.

After the paper had been published, he had received two visitors. One was Clerc, who had brought the disturbing news of the disaster at the Lavenière estate. Clerc had suggested that Andréus publish the news in his next issue, along with the advice that all but essential citizens should leave the town for Fort-de-France. Hurard had demurred, taking refuge in the argument that only the Governor could issue such an order. But Louis Mouttet, appointed Governor just seven months earlier, was determined that nothing would disrupt the island's elections, only nine days away.

Shortly after Clerc had left the newspaper office, Thomas Prentiss, the American Consul, had called with the news that the captains of two of the American merchantmen in the roadstead were weighing anchor because they believed the island was in danger. They had taken on board a number of families from the American colony of St. Pierre.

Hurard had said he would publish this item as an example of unnecessary panic.

"That, sir, is a matter for you," Prentiss had retorted, "but you may well come to regret it."

Alone in his office Hurard had begun to wonder whether there might not well be some truth in what the diplomat had said. As he stood there pondering the events of the day, clear across the town had come a new boom from the throat of Mount Pelée. It had been followed by a fresh fall of ash that had blotted out the sun and driven the citizens to take shelter once more. He wondered again how long he could continue with his deception in return for the Governor's favors. For those favors, he would lend total political editorial support in the election campaign.

So far Hurard had honored his side of the bargain. Since the election campaign had opened three weeks before—coinciding, in fact, with the first signs of life in Pelée—*Les Colonies* had devoted massive editorial space to supporting the manifesto of the Progressive Party.

Reactionary in its outlook, its roots buried deep in traditional French colonial policy, the Progressive Party stood for total white supremacy. For centuries, with its cohort of political die-hards, the Progressive Party had almost automatically produced the senator and the two deputies, one from each of the island's *arrondissements*, who represented the island's political views in Paris.

Then at the 1899 elections three years before, the Radical Party—the voice of the island's politically emerging Negro and mulatto population—had come from nowhere to win a spectacular victory at the polls. Their candidate for senator, Amédee Knight, a forty-seven-year-old Negro, had been elected. Nobody had been more surprised than the happy-go-lucky policy-makers of the Radical Party. But in Knight they had an outstanding candidate. Educated in Paris at the École Centrale, versed in the ways of white thinking, Knight had proved formidable, first as an assistant mayor of Fort-de-France, then as secretary of the town's Chamber of Commerce, and later as president of the Town Council. For the first time the black population had a voice in Paris; and while in many cases that voice was listened to with polite patience, Knight was able to push for a number of reforms in education, housing, and employment. In Martinique itself he played a crucial part in transforming the Radical Party into a purposeful machine. Now, in the elections of 1902, the Radicals were making a firm bid to wrest total political control of the island.

Opposing these Radicals was Fernand Clerc. The events which had brought the island's most distinguished planter into the election ring were simple enough. Shocked by Knight's election as the island's senator, the Progressive Party had realized that it would need a more liberal image if it were not to be defeated when the island next went to the polls on Sunday, May 4, 1902. After considerable persuasion, Fernand Clerc, who paid only lip service to the Progressive policies, had been persuaded to stand for the northern *arrondissement*. A generous employer, a man who had made almost a fetish of good labor relations, and deeply religious, Clerc was outwardly the ideal choice to represent the Progressive Party's new political approach.

But already there were those in the Party, Hurard among them, who were regretting their choice. The Progressive Party stood on a simple political platform: for decades it had controlled the political fortunes of the island with cause for little complaint—why change now? The message was implicit in every piece of official propaganda disseminated by its efficient Party machine. But Clerc had moved well beyond the carefully prepared Party brief. He had admitted the need for reform in a number of areas where white influence dominated, such as in labor relations among the workers of St. Pierre and in some of the smaller sugar factories outside the town. He had called for stricter control over prostitution and the town's night life. He had urged a return to a more godly life.

The result had been extraordinary. A large section of the white population had been first aghast, and then disgusted with Clerc: many of them had made fortunes out of exploiting cheap labor, and running brothels and back-alley hotels. To give up all this would be unthinkable. *Les Colonies* had been bombarded with irate letters from white readers. So far not one had been published: politics by suppression was nothing new to Andréus Hurard. His decision not to publish had been reinforced by Governor Mouttet. On the other hand Clerc's policies had been received with suspician by the Radical Party; to them it was a clever electioneering stunt: empty promises that would be forgotten once Clerc had been put into office.

In three short weeks, Fernand Clerc, politically naive almost to the verge of incredibility, had succeeded in doing considerable damage to his own stature, alienating a large portion of his vote, and almost guaranteeing that the Progressive Party would be torpedoed at the elections. He had become a political maverick.

The one man who could have stopped him from his headstrong, lonely course was Governor Louis Mouttet, an open supporter of the Progressive Party, even though as the appointed representative of the French Government he was theoretically under orders to keep out of party politics. But from the time Mouttet had arrived on the island with his wife and family seven months before, relations between him and Fernand Clerc had deteriorated from chilling politeness to icy indifference. Mouttet, self-made, pulling himself up by his political bootlaces, stood for everything that Clerc, the gentleman planter, found distasteful. Mouttet and a few others had made no secret of the fact that they opposed the choice of Clerc as candidate. In an election already fraught with difficulties, these personality clashes would play a decisive role.

They would also have a fatal bearing on the effect Pelée would have on St. Pierre's voters.

On the far slope of Pelée, out of sight of St. Pierre and the falling ash, a party of villagers was making its way up a narrow mountain trail. The group was led by an anxious bridegroom who was searching for his bride-to-be and the old voodoo priestess who was to have been her guide to the wedding, now delayed, in the village below.

The bride had been receiving ancient voodoo instructions in the art of love and the customs of marriage. Her mentor was one of a handfull of specialists on Martinique skilled in the art of preparing young women for love, and

The Anxious Bridegroom

the girl had spent the previous week at the old woman's cave, being initiated. Only a few hundred yards from the cave's mouth, the bridegroom and his party found the young girl in her bridal gown and her guide, their bodies hideously scalded. They lay beside the steam vent which must have burst by the trail just as they passed.

CHAPTER THREE

A Cell with a View

FATHER ALTE ROCHE, standing on the peak of Mount Verte, two miles to the south of St. Pierre, viewed the volcano with professional detachment. The tall, bony Jesuit had spent his lifetime studying the cause and effect of volcanic disturbances in the Caribbean. For the past three weeks he had been carefully observing and recording the behavior of Pelée. Every few days he had picked his way over the pumice slopes of Mount Verte to stand on its summit a thousand feet above sea level, a perch that offered him a clear view of Pelée. A cautious man, he had noted his observations in a log book, keeping them to himself, trying to assess what he had seen by what he already knew about the volcano.

Its geology he knew well. Pelée, like the other four hundred mountain peaks on Martinique, was composed almost entirely of volcanic material that had been thrust up through fractures in the sea bed when the earth's crust was cooling. Pelée itself, one of the five hundred volcanoes that have ever been active in historic memory, lay in the center of one of the great volcanic belts that circle the oceans of the world.

The largest of these is the ring that surrounds the Pacific. Starting in the South Shetland Islands, several hundred miles south of Cape Horn, the chain extends up the west coast of South America, across Central America and on up through the North American continent. In Alaska it crosses the Pacific to the Aleutian Islands. From there it sweeps on to the eastern seaboard of the Pacific, taking in Japan, Formosa, the Philippines, the Solomon Islands, and New Zealand, before finally ending on Mount Terror on the Antarctic continent. In all, the chain stretches for nearly twenty-five thousand miles, encircling over two hundred million people.

Eastward from the Pacific, another belt extends through Sumatra and Java. It is the most active of all the chains. Here over one hundred volcanoes constantly display nature's energy. Here the natives live beside a gigantic boiler that bubbles and burbles and regularly spews out liquid rock and gases from the very core of the earth.

The Atlantic is edged on three sides with a volcanic chain. It starts in the far north, traveling down through the peaks of the Azores, the Canary Islands, the Cape Verde Islands, Ascension, St. Helena, and Tristan da Cunha. On the western side of the ocean are the volcanoes of the West Indies. They include the peaks of Saba in the north, rising nearly three thousand feet above sea level, to Grenada in the South, from whose peak on a clear day the outline of the South American coast is visible. In the center of this Caribbean branch of the great Atlantic belt is Pelée.

Now, late in the afternoon, Pelée belched again, spewing hot dust as it cleared its throat.

From his position, Father Roche looked out across the intervening molder of jungle vegetation to the volcano. Below him the ground was a patchwork of swamps, streams, and small rivers. Sprinkled among them were plantations and hamlets. In the middle of the patchwork was the Guérin sugar refining factory, one of the finest on the island. Its tall chimney stack poked its way eighty feet above the factory buildings. Inside those buildings 148 workers processed sugar cane into crude sugar, which was then taken by carts to St. Pierre and shipped to America and Europe. Beyond the factory lay the lower slopes of Pelée, hidden beneath a thick coat of rotting green. This covering rose to a deep collar of scrub-covered lava rock. Above the collar stretched the neck of Pelée, short and massive, ending in the open mouth of the crater, which had yawned again to discharge ash.

The wind had carried the ash falls clear of Pelée's neck and collar to deposits that began at a point halfway down the green coat and spread far and wide from there.

From the top of Mount Verte, Father Roche had observed after one of his previous visits that "the scene is one of beauty. Up here the air is free of sulphur fumes, and the effect is similar to looking down upon a land covered with a gentle frost through which the varied colors of nature are reflected." Father Roche watched the latest layer of gentle frost drift lazily down on the countryside. He noted that it "appeared to be no different from all the other falls that have been ejected these past days."

The volcano fell silent, save for a strange humming and vibrating noise that carried clearly across the intervening ground to where the priest stood contemplating the mountain.

He had come here for a better view of the mountain for a special reason this day. His parish was in the mulatto quarter of St. Pierre. There rumor often ran rife, spead by superstition, excitement, or fear. So far Father Roche had detected little alarm among his parishioners because of Pelée's awakening. But he reasoned that after the heavy falls of ash that day, the time would be approaching when he would be called upon to offer reassurance.

Already one of his parishioners had confided after morning Mass that she believed that "God's blacksmith was hard at work in his forge," and that the sparks from his anvil were escaping through Pelée. Even after a lifetime in the Church, the Jesuit found it hard to convince his parishioners that pagan beliefs and true religion should not go hand in hand, as they did on Martinique. He had explained to her that there was little to fear, adding that if there were, his own experience would enable him to give ample warning. He, like others on the island, based his opinion on the facts as he saw them. He discounted the humming and vibrating that carried across to his perch as being of any significance, except that he felt they could be the grumbling of the volcano settling down again. He could not be blamed for his conclusion. His world had never known a volcano to behave as Pelée was to behave.

Outwardly, the volcano offered the traditional conical

shape of volcanoes the world over. In the last twenty-four days she had acted like any other volcano stirring from sleep: ejecting dust in a sudden uprush of gas and steam, and at night lighting up the sky with brilliant flashes. There had been, for Father Roche, something majestic in this natural spectacle. It was a visible reminder, in his eyes, of the infinite power of God.

He knew, from personal observations and from his reading, that there were a number of different kinds of volcanoes.

Those of Hawaii—Mauna Loa and Kilauea—had given their names to the most spectacular kind of eruption. When they erupt, they send great fountains of debris cascading into the sky. Mauna Loa, the largest volcano in the world, rises from her seventy-mile-wide foundation to glower nearly fourteen thousand feet above the level of the sea. On the rampage, her throat glows, fanned by the mighty bellows working away on the ocean bed. Around her peaks miniature tornadoes stalk, fanned to life by the intense heat pouring out of the crater.

Father Roche remembered seeing, as a young man, another type of volcanic eruption. From time to time Stromboli would light up the Sicilian landscape where he had been born fifty years before. Stromboli, a natural lighthouse familiar to the Mediterranean traveler for centuries, had been glowing since the dawn of history. Its discharge was slight, similar in fact to the way Pelée had been discharging.

Other kinds of eruption were known. The *vulcanian* type is explosive and sudden, caused when a plug of lava that has blocked a volcano's throat is burst by the pressures beneath and a mass of solid and liquid rock is hurtled into the area in great clouds of vapor and dust. The *Icelandic* kind of eruption occurs when billions of tons of fluid lava pour steadily over thousands of square miles of countryside and sea, wasting everything it touches.

Finally, Father Roche knew of the kind of eruption that took its name from Solfatura in Italy. The crater there last ejected lava in the twelfth century. Since then it has discharged only gases and an occasional trace of ash. Now looking across to Pelée, Father Roche wondered whether its crater, silent and indistinct in the fading light, was not the *Solfaturic* kind, constantly threatening but never quite living up to its promise of impending violence.

All he knew now, as he picked his way down the slopes of Mount Verte to make his way back to St. Pierre, was that he would be able to offer reassurance to those who wanted it: the volcano, he decided, was still not a threat. Even the vibrations and humming had died away. He felt he had been right to discount their importance.

He was not to know that Pelée would give its name to a new kind of eruption—one in which the humming and vibrating would have a part, one more incredible than the world had ever known.

The Governor of Martinique, Louis Mouttet, had come to a decision: he would do nothing more. That day he had authorized a number of actions. He was well satisfied with the results they had achieved. Now, in the darkness of tropical night, he stood on the balcony of his Residency in Fort-de-France, reviewing in his mind those results.

All that day he had demonstrated his power. It had been early morning when the first report of Pelée's latest behavior had reached him. It was the news that the Lavenière estate had been isolated by lava and that there had been loss of life. He had dispatched a troop of soldiers. They had reached the estate by descending the side of the valley on which the Lavenière villa was built. They had returned with the news that Mademoiselle Lavenière was in a state of shock after witnessing her father's death but that otherwise the estate workers appeared calm. Mouttet had then ordered a doctor to be sent from the military hospital in St. Pierre to attend to Suzette Lavenière.

Later, there had been reports from the town itself. He had been told that the ash fall was heavy and that by midday a number of animals and birds had been choked to death by the sulphur fumes. Soldiers from the St. Pierre garrison were ordered to remove the carcasses fom the steets and bury them outside the town.

Then had come the news that the American consul, Prentiss, had been troublesome. The diplomat had, after

A Cell with a View

his interview with Hurard, the editor of *Les Colonies*, tried to telegraph an alarming report to Washington. It spoke of the need "to be on the alert" for a possible eruption from Pelée. There had followed what in Mouttet's eyes was nothing more than hysteria: "The rain of ashes never ceases. The passing of carriages is no longer heard in the streets. The wheels are muffled. Puffs of wind sweep the ashes from the roofs and awning and blow them into rooms of which the windows have imprudently been left open. . . ."

There had been more in this vein; however, the St. Pierre telegraph office had lost its overseas connections. When the message was sent to Fort-de-France for relay to America, Mouttet was shown a copy, and the Governor gave strict instructions that under no circumstances was the message to be sent. He then sat down and wrote Prentiss a curt note, pointing out that it had come to his attention that "you appear to be spreading alarm abroad, and that by doing so you could create a state of false fear and pessimism where none need appear." He had sent an aide to deliver his note to Prentiss in St. Pierre. So far he had received no reply.

For a while life had returned to normal at the Residency. Lunch had been served, as it was every day, on the balcony. There Mouttet and his wife, Maria, had dined leisurely, enjoying the panorama of town life that revolved below them.

Madame Mouttet was one of those inexplicable Frenchwomen who contrive to seem, in spite of a moderate figure and ordinary brains, to be almost unattainably elegant and intelligent. On the other hand Mouttet showed signs of personal and tropical ravages. A fondness for food and wine had thickened his body while dysentary and malaria had coarsened his features.

He was now coming to the end of his seventh month in office, this gross Frenchman of forty-five. He knew that to return to his native Marseilles—four thousand miles and six weeks sailing away—would be to return to a life doomed to end in boredom. He was unlikely to find a position of any note in the Foreign Ministry in Paris; he was unqualified for a niche in the commercial or industrial life of France. In Martinique he was titular head of a kingdom that he ruled with an indolent hand. He was determined to go on ruling until he died.

He brooked no interference from outside, and the island presented no political problems for the French Government. It was a situation of mutual satisfaction.

Born the son of a farm laborer in Marseilles on October 10, 1857, he had overcome the drawback of a rudimentary education to admit himself to a local businessmen's group. Later, in 1883, when the Cercle Saint-Simon moved to Paris, Mouttet went with them. Through the Cercle, a form of freemasonry, Mouttet got himself a job as a proofreader for a weekly newspaper, *La Patrie*, in the Rue d'Aboukir. Thirty years later his editor, Emile Massard, was to remember the ambitious young Mouttet: "He used to work until 2 A.M., sometimes going without dinner, kept going by a single dream: that of getting a byline on the law report page, a job which consisted of cutting the reports out of the evening papers and adding a signature. One day, by chance, his dream came true. He was asked to deputize for an absent colleague, got his fingers into the scissors, and never took them out again. From that time Mouttet was a happy man. He bought himself a top hat and started to frequent Paris café society."

In the politically aware society of the smart Parisian cafés, Louis Mouttet developed a taste for power. In no time at all he had decided to give up journalism, and again the Cercle Saint-Simon came to his aid, finding him a job as secretary to the Historical Society of Paris. It was a post that paid well and allowed him ample time to pursue his active social life. The young Mouttet cut a dashing figure in his opera cloak, going from one salon to another. Women found him attractive. Blue-eyed, with curly black hair and a well-trimmed goatee, he was the darling of the boulevards.

In 1886, his ambitions taking wing, he entered the French Colonial Service. But more than good looks were needed to progress in the Colonial Service. In three years, in spite of all his connections, Louis Mouttet had managed to be no better than personal secretary to the Governor of Indo-China. But the post gave him a taste for the colonial

A Cell with a View

way of life. He was determined that no matter what happened, he would never give up the retinue of servants, the long siestas, the leisurely evening highballs that, the world over, have been the hallmarks of colonialism. He needed one more prop to make his life complete—a wife.

In the spring of 1890, on leave from his latest post as Director of Internal Affairs in Guadeloupe, he met Maria Coppet, the niece of the deputy for Le Havre. In September of that year he married her, and the same day the French Government awarded him the cross of the Legion of Honor for "political services in the name of the Republic." From then on, helped by Maria's family, Mouttet's career in the Service advanced rapidly. After a short spell as Director of Internal Affairs in Senegal, he became Interim Governor of the Gold Coast in 1895. A year later he was made Governor Fourth Class; in May, 1898, he was promoted to Governor Third Class. The following year he became Governor of French Guinea and became briefly known to the world as the man who actually got Dreyfus to return to stand trial in Paris. Shortly afterward he was made Governor Second Class, and in November, 1901, he came to Martinique with his wife and three children to administer the island for France.

He had fallen in love with St. Pierre. Mouttet and the town deserved each other. Raffish, conniving, and corrupt, they recognized a mutual need in each other, the need to live and let live.

Mouttet had let it be known to the island's voters that he believed they were entitled to live their lives by their own lights, providing they did nothing to sully the good name of France. In supporting the Progressive Party, he was supporting a platform that was a license to carry out political and social mayhem in St. Pierre and to a lesser degree in Fort-de-France. It was a license that Mouttet was willing to endorse in exchange for an easy Governorship.

But there was Pelée. When the volcano had first stirred, Mouttet, with the animal sense of a born politician, had recognized its potential menace. If it continued to grow and threaten St. Pierre, there was likelihood of panic.

Panic would cost the Progressive Party further precious votes.

He had already taken out the most effective insurance he knew against this possibility by increasing his influence over the island's leading newspaper, *Les Colonies*. For years the paper had faithfully supported the ruling Governor on every issue he raised. In return, a succession of Governors had exerted their own pressures on the business community of Martinique to help finance the newpaper by advertising. It was a crude but effective piece of press manipulation. For Hurard, the editor and publisher, it represented a guaranteed income in return for following a line he was personally not against.

When Pelée had started grumbling, Mouttet had told the editor that it would be in everybody's interest if the paper dismissed the possibility of any threat from the volcano. In a piece of simple horse-trading, Mouttet had argued that unrest would create a decline in the advertising revenue, as well as affect the electoral attitude. In the face of this threat, editor and Governor had closed ranks completely to present, through *Les Colonies*, a dangerously distorted picture. Regularly the paper's editorials stated that there was nothing to fear.

But that afternoon had come a piece of news that had provoked concern even in Mouttet's mind.

It had been brought by Sequin, the master of the *Topaz*. He had brought the news that Le Prêcheur "appears to be covered in ash which has set the houses alight. Lava flows down to the waterfront and into the sea, sending clouds of steam rising."

Sequin had been too far away from the shore for his passengers and crew to get a clear view of the devastation, but through his telescope it had shown up clearly. Having no passengers or cargo to unload at St. Pierre, he had steamed straight to Fort-de-France to make a personal report to the Governor.

Mouttet had counseled Sequin to keep what he had seen to himself, and to take the *Topaz* on to St. Lucia. There, the captain was to make "discreet inquiries" as to what had happened to the telegraph link between the islands. Mouttet had expressed the opinion that it was prob-

A Cell with a View

ably no more than a simple technical fault. It would be some time before he learned the truth.

Late that afternoon Sequin steamed from Fort-de-France on his mission for the Governor. In his Residency Mouttet reviewed what the sailor had told him. Earlier that day he had ordered the military commandant at St. Pierre to quarantine the refugees from Le Prêcheur to prevent panic. He concluded that he had been right to confine the villagers: if any of them had returned to their village and found it in the state that Sequin described, they would undoubtedly have spread alarm far and wide.

Certainly, the situation in St. Pierre itself seemed to be well under control. In his diary for the day, Mouttet wrote: "In all, they present a picture of a town coping well with a trying situation. Any signs of panic have been quelled."

He ignored the report that had come from the town of a cobbler who was behaving oddly. The man had spent the day in his shop in the Place Bertin refusing to open the door to anybody, shouting out to callers that "doom is at hand." Soon the man had become an object of fun, and Mouttet, like any politician, recognized the value of ridicule as a weapon to cloud more important issues.

He had also ignored the two requests Fernand Clerc had sent to the Residency by messengers. Both had asked Mouttet to visit St. Pierre to "view the situation for yourself."

But Mouttet refused to make a special trip. He had to go to St. Pierre on the eve of Ascension Day anyway for the Mayor's annual banquet. Together, he and his wife would drive in state through the town. It would be an effective demonstration to the citizens of what he thought of any threat from Pelée.

In St. Pierre that evening Louis Auguste Ciparis didn't feel like eating. He was in the condemned cell of the massive military prison in the northern part of town. He was due to be hanged in less than a week's time, on Thursday, May 8, Ascension Day. Now, his body spent and broken by weeks of brutality from the guards, he lay on a straw sack in his dark prison cell and listened for a sound in the night. It would be the sound of workmen dragging timber across the prison yard. The wood would be used to build the scaffold on which Ciparis was scheduled to die. Every day and night since he had been condemned to death, he had listened for the sound. And every night he had fallen asleep exhausted from the strain of listening.

Tonight, May 2, 1902, would be no different. He was fast asleep when the workmen brought the timber. They carried it on their shoulders so as not to disturb the prisoners. Their footfalls were deadened by the thick layer of ash. They quietly deposited the wood in a far corner of the prison yard and returned the way they had come.

SATURDAY

❧

May 3, 1902

CHAPTER FOUR

Tumbled from Bed

SATURDAY, MAY 3, produced one of those spectacular dawns that God reserves for the tropics. On the eastern side of Mount Pelée a party of men were preparing to set out again. For three days they had steadily quartered the area, working their way ever upward toward the volcano's pumice collar. They numbered fifteen in all: tall, supple-limbed men, barefooted, clad in the *jupee*, the traditional smocklike garment of the island.

The behavior of Mount Pelée had produced a unique reaction in them. They were positively elated by its grumbling and coughing. The noise had stampeded the wild pigs that roamed the upper reaches of Pelée on to the muzzles of their muskets.

The prevailing wind had insured that little ash had fallen on their side of the mountain. It had also carried away from their earshot most of the throat-clearing that Pelée had indulged in; from their position, a thousand feet below the summit, the noise had resembled little more than the sound of a rum cask being rolled along the water front at St. Pierre. And they had not heard even that much during the night, although they had slept lightly, ever on the alert for the sounds of a wild pig on the rampage.

They had no doubt that the noise had been responsible for driving the boars from the other side of the mountain. It had provided the hunting party with a rich haul. In three days they had killed twenty-four pigs, the largest number any of them could recall ever having been captured in one hunting safari on the island. They had cooked and skinned the boars, and wrapped the flesh in leaves and sacking. Later they would sell it in the mulatto market in St. Pierre.

Rich though their catch had been, they were loathe to leave the mountain. They knew from experience that the largest hogs lived among the scrub of the volcano's collar.

They decided to sweep across a sector of the collar, work their way around to the far side of the mountain, and then descend its slope and make their way into St. Pierre. They would hunt as they went.

Among them they carried not an ounce of fat on their bodies. A lifetime of climbing the island's mountains and living on a simple diet had achieved this. Lean, tough, and durable, having eaten they could march or hunt for hours. They had breakfasted on corn, cassava bread, and coffee brewed in an iron skillet. As the sky lightened, they set off, carrying their guns and machetes. Behind came the baggage boys, carrying the bundles of baked pork, cooking utensils, and spare weapons.

They made their way up through the solidified lava, moving in a broad diagonal from right to left. The ground was pitted with great gaping funnel-shaped holes caused by previous eruptions, holes down which a man could plunge to his death. But it was up here that the fattest and fiercest of the wild pigs roamed.

By the time the sun was full in the sky, the hunting party was working its way around the edge of the collar. Below them stretched the western side of the island with St. Pierre itself drawing back from the sea to disappear a short distance inland.

For a long time the hunting party looked at the scene in stunned silence.

The whole area appeared to be caught in the grip of deep winter; from a thousand feet below where they stood, as far as their eyes could see, everything was white. From beyond Le Prêcheur on one side, the whiteness extended down to Gele Peak, seven miles away, then curved in a broad arc to cover the sea. St. Pierre, roughly in the center of this area, looked to the hunting party as if it had been subjected to particularly heavy snowfalls.

Immediately below them, a few hundred feet away, was the source of the Roxelane River, which journeyed down the side of Pelée, flowing through the valleys which dropped sharply from its slopes and on into St. Pierre. Normally the Roxelane began life as a trickle seeping out of the volcano's side, gathering strength as it tumbled, to course briskly in the rainy season through the town and out to sea. The trickle had become a fast-flowing stream, pouring from a number of cracks to unite and flow at speed, swelling the Roxelane to twice its normal size. Below them in the valleys they could see that the river had breeched its banks in several places, but that in doing so it had considerably dissipated its force, and there was no danger, yet, of flooding in St. Pierre.

How long that situation would remain appeared debatable to the hunters. A number of other streams were also in flood. The Blanche, which flowed into the sea beside the Guérin sugar factory, was roaring down the mountain side; the La Mare, usually sluggish except when the rainy season was at its peak, was now rushing swiftly down the side of Pelée and into the sea north of St. Pierre.

Below them, from La Calebasse—a hump of solidified lava growing out of the side of Pelée's neck—stretched a flow of thick brown mud that had rolled down the volcano's side and into a tributary of the Capot River, three thousand feet below. It had forced its way down the tributary and surrounded the Lavenière estate, effectively sealing it off from three sides.

Apart from the gushing of the water from the fractures in the collar, there was silence.

From this vantage point nearly three thousand feet above the level of the sea, the tiny toy ships riding at anchor in the roadstead off St. Pierre appeared to be settled on a sheet of white.

Then, from the northern end of the town, the hunters saw a ragged column of tiny black specks making their way along the road to Le Prêcheur. Apart from them, nothing else moved. From the chimney stack of the Guérin sugar refining factory in a valley below, a wisp of smoke rose into the still air.

The hunters stood there contemplating the scene,

wondering at what they saw, unable to believe that the crater, brooding silently above and below them, could have been responsible for so dramatic a change in the landscape.

They were still standing there when they heard a deep panting. It was still some way off, but approaching steadily up through the scrubland. Swiftly the hunting party dispersed, taking cover behind the rocks, waiting for the wild pig to get closer. Soon they could hear him clearly, trotting and dislodging small stones in his progress upward.

The hunters got ready to meet his charge once he had scented them. He had already started to come across the ground very fast; he was huge, fierce, his curving tusks sweeping ahead of him, seeking a target. He charged up to the hunting party. They scattered, taking care that they were not in each other's crossfire. As the boar reached the rocks where they had sheltered, it wheeled and circled, snout down, seeking a victim. One of the hunters fired. The bullet glanced off the hog's hide, and he turned and raced back the way he had come, his resolve to stand and fight gone. Behind him, the hunters started to leap across the rocks in pursuit, firing as they went.

The animal was halfway down the volcano's lava collar when the ground appeared to dissolve beneath him. Where one moment there had been firm rock, there was now a bubbling brown liquid mass; even from a distance of several yards the hunters could feel the heat it gave off. They watched the hog, screaming in terror, being sucked into the morass. Then the liquid belched upward, carrying the dying pig with it as it started to seep down the mountainside like thick, hot molasses.

The hunters turned and fled the way they had come. Behind them the hole in the ground broadened its perimeter until it was roughly thirty feet across. When it had reached this size, the lava stopped draining from this new abscess almost as quickly as it had started.

The sun was rising behind Pelée, tinting its peak a reddish brown, when the refugees of Le Prêcheur left the outskirts of St. Pierre. They moved quickly, as if they were eager to be on their way, the adults carrying their small children, the older children almost running to keep up.

Few of them had slept that night, herded, as they had been, into the confines of the Town Hall compound. The night had been warm and still and, for them, long. Through the hours of darkness, they had talked to and fro among themselves, these simple fisherfolk and market gardeners, about the situation they were in. In the early hours of the morning, they had come to a decision; they had decided that they could not accept the garrison commander's order that they must be kept confined. They were country people, used to freedom; to have it denied them, as it had been late yesterday afternoon when soldiers had shepherded them from the Place Bertin, was something they would not tolerate.

They would return to Le Prêcheur. Their parish priest had gone to the compound gate and awakened the soldier guarding it. The priest had told him his parishioners planned to return to their homes.

The soldier had said they would have to get permission from the garrison commander. The priest, a bold man, had ordered the guard to go at once, in the name of God, to his commander and inform him of the villagers' wish. The soldier had gone, confused in the face of the authority of the Church, to the garrison headquarters on the far side of town. There, a duty officer had sent the soldier back to the compound, forbidding him to let anyone leave until the garrison commander awoke in two hours' time. When the guard returned to the compound, he had found it empty. Now, as he made his way miserably back to the headquarters to report the loss, the priest was leading his flock back down the road to Le Prêcheur.

He walked ahead of his parishioners, a lone figure in an ash-stained cassock. He was uneasy about the course of action he had endorsed.

The priest knew that the villagers must have been rounded up on the instructions of Governor Mouttet. He knew, too, that Louis Mouttet would not have held them in custody unless he had good reason to do so; the only reason the priest could think of was that the Governor had decided it would affect the Progressive Party's chances of being re-elected to office.

Panic, reasoned the priest, would affect that chance. Yet he had seen nothing in St. Pierre before the troops came to take them to the compound that suggested any reason for panic. In fact, when some of the villagers had tried to describe what had happened to their village, the townspeople who had bothered to stop and listen had looked upon them as if they were stupid country folk who had panicked without good reason.

Therefore, concluded the priest, if the cause of the panic was not in the town, it must be somewhere in the countryside; somewhere, he sensed, without knowing why, could be in the region of Le Prêcheur.

This morning, with the sun up, the countryside silent, he seriously debated whether to turn his flock back to St. Pierre. To him, the land had suddenly taken on a sinister aspect. On either side of the road were heaps of freshly dug earth. They were the graves of the animals from the town who had suffocated under the ash fall. He looked to his right across the rising ground, up toward Pelée, massive in the sharp light, and he wondered again whether he had made the right decision.

Then something happened that resolved the matter for him. From behind came the sound of singing. The sound came lightly at first, then grew in strength as they turned a bend in the road and St. Pierre could be seen no more, the proud notes filling the air.

They were the words of the Lord's Prayer.

The parish priest, his own eyes red from the emotional impact of the moment, turned and looked back on his flock. He felt he knew at that moment what Moses must have experienced when he had led his people out of Egypt.

James Japp, the British Government's Consul General in Martinique and Guadeloupe and their Dependencies, was awakened this morning by a sudden surge of water which sounded as if it were rushing beneath his bedroom window. He looked at his bedside clock. In the dim light of the room, he saw that the hands were coming up to six o'clock. He could not remember when he had been awake so early before. For a moment he listened, staring at the ceiling. Nothing disturbed the silence of the house except the steady rush of water. He rose from his bed and made his way to the tightly shuttered window. He pushed open the shutters and looked out on an astonishing sight.

The British Residency was situated on the outskirts of St. Pierre, standing on a mound around which the Roxelane River flowed. The over-all impression of a castle and its moat was heightened by the two footbridges that linked the Residency with the far banks of the river, and the sturdy stone walls that gave it a curiously medieval look even with the tropical vegetation that clung to the stone. The impression of an Imperial fortress was one that James Japp delighted in fostering. To him the residency was a bastion of all that was great in Britian.

But now as he stood there wide awake in the bright hard light of early morning, James Japp suddenly felt very vulnerable. In the fifteen years, two weeks, and three days he had lived in the Residency, he had never known it to be threatened as it was being threatened now.

The Roxelane River had overflowed its banks, and the flood water was lapping against the very outer walls of the Residency. Where the river bed had been, slow and sluggish in its passage to the sea at this point of its journey, there was a fast-flowing tide of muddy water carrying dead farm animals, vegetation, and trees.

The flood water had completely isolated the Residency, except for one link. James Japp had never approved of that link. He regarded the telephone as a monstrous intrusion into all levels of life as he knew it.

Japp, a courtly, kindly man, was cast from the mold of late Victorian diplomacy that helped Britain maintain her hold on the world. It was a mold where good manners and breeding were of paramount importance and where panic, indecision, or haste had no room.

Calm and reason had stood him in good stead all his life. He had come to the island on May 11, 1887, as British Vice-Consul. He had never been an ambitious man; he had often said that "to serve Her Majesty" was enough, no matter where the service was. He had known what it was to have money all his life; it had cushioned him against all constrictions of Victorian life, and it had counted heavily in his favor when he had been accepted into the Consular Service. During the years he had served Her Majesty's Government, in one quiet capacity or another, he had found that money had been a useful buffer enabling him to live at a high standard no matter where in the world he was stationed.

In Martinique he was quite aware that he had established for himself the role of *doyen* of the diplomatic corps, a position that had been maintained from that day, September 6, 1897, when he had been elevated to the rank of Consul General at the age of forty-one.

Now, in his forty-sixth year, James Japp sensed, and was unable to find an explanation for it, that the swirling water was a precursor of destruction more terrible than he could ever imagine.

He turned from the window and walked across the bedroom and down the stairs to where the telephone stood on its stand. For a moment he looked at it, a nineteenth-century man reluctant to become involved with progress. He picked it up and waited to be connected to the number of the American Consul, Thomas Prentiss, who lived in L'Centre, the accepted diplomatic quarter of St. Pierre. It was a few moments before James Japp realized that the line was dead.

From that moment the British Government's representative in Martinique was completely cut off from the outside world. There was only one sensible course of immediate action left open to a man like James Japp.

He picked up a small hand bell beside the telephone and shook it briskly. From the back of the house, a white-coated manservant appeared, surprised to find his master standing beside the telephone in a striped nightshirt.

"I want breakfast early today, Boverat. I want two eggs, three rashers of bacon, and toast. And tea."

With mounting disbelief Thomas Prentiss read to his wife, Clara, the main front page story of the day's issue of *Les Colonies*. It was headlined: "On to Mount Pelée." There followed the single column of type that had caused

Prentiss to react as he did. It stated that the excursion to the volcano which the paper had referred to in yesterday's issue would still take place on the following day, Sunday, May 4.

Prentiss read: "Those who have never enjoyed the panorama offered to the view of the astonished spectator at a height of more than four thousand feet, those who desire to see close at hand the still yawning hole, should profit by this fine opportunity and register their names this very evening at the latest. . . ."

There followed detailed instructions on the time and route of *la grande excursion*. It would start at three-thirty precisely the following morning from Marche du Fort, a residential quarter of the town. After describing the route, the story ended with these words:

"Those who do not care to trouble themselves with food should pay an assessment of three francs. They will not regret being relieved of the trouble of procuring food. To judge from the list of those who are going, the company will be very numerous, and if, therefore, the weather be fine, the excursionists will pass a day that they will long keep in pleasant memory."

Thomas Prentiss, a man not easily shaken, put aside the newspaper and looked at his wife.

"I think," he said, picking his words carefully, "I think the whole world is going mad."

In his St. Pierre office in the Rue Victor Hugo, Fernand Clerc had come to the same conclusion. From a number of his estates dotted around the foothills of Pelée had come reports of flooding, crops killed by the fallout of ash, animals dying from the sulphur fumes. And yet here in St. Pierre, nobody seemed to care.

But Clerc was determined to make them care. In an hour's time he would hold a meeting in his office. He had asked a number of important businessmen, as well as Father Alte Roche, to attend.

Now, as he read the latest edition of *Les Colonies*, he could not help but wonder at the sickening dishonesty of politics.

Louis Mouttet could only admire the skill that Hurard had put into the newspaper. There was a swinging attack on the American families who had "scuttled" at the sight of a little ash and a biting story about a foreign diplomat spreading alarm abroad which contained the substance of the cable that Prentiss had tried to send. But what gave Mouttet particular pleasure was the main editorial.

In four hundred carefully chosen words of invective, the Radical Party had been politically assassinated. They were accused of creating panic where none need exist.

And why had they done so, demanded the editorials; why, for no other reason than that they hoped it would gain them office where they could practice their "racialist policies."

Those two words brought out into the open the political issue that had been simmering during the entire election campaign.

The editorial reminded its readers that the occupant of the condemned cell in St. Pierre's military prison, Ciparis, was a Negro; it also reminded them that he had killed a Frenchman. "Was that to be the policy for the town if the Radicals come to office?" demanded the editorial. "We have heard much talk about racialism from our opponents. But are they not guilty of racialism by agitating for the release of his Negro for so foul a crime?"

The fuse had been lit. Now it could be kept alive until polling day, with *Les Colonies* ever shortening it; Hurard could, and indeed would, return several times to this theme in the coming days.

Governor Mouttet particularly admired the way the editorial thundered to a close: "So when you hear tales of panic and impending doom to St. Pierre because ash is falling from Mount Pelée, weight those tales carefully. They may have been spread by Radicals eager to drive you away from your duty—to oppose them at the polls!"

The newspaper had fired a broadside which in the coming days would have a decisive effect on events in St. Pierre. Its final effect was to be of far more consequence than its influence on the political results of a Martinique election.

Tumbled from Bed 49

Suzette Lavenière had come to a decision after lying awake for many hours: she must, somehow, escape from the room that now contained her and make her way back to the estate. In those hours she had tried to recall the events which had brought her to a hospital cot in the military hospital in St. Pierre. Her recollection was hazy and filled with gaps. All she could remember clearly was a doctor and a nurse coming the previous afternoon to the villa. The doctor had looked at her briefly and murmured a few words to the nurse, and shortly afterward she had been transferred from her bed to a makeshift stretcher, fastened to it, and caried up the valleyside by two of the estate workmen.

Once she remembered looking back. Mount Pelée, which had suffocated her father with a mouthful of mud, filled her vision. She screamed, and the doctor had given her a bitter potion from a bottle. Then she had fallen asleep.

When she awakened, she found herself in darkness in this iron bedstead. And when dawn came on this Saturday, Suzette saw that she was in a small room, its walls whitewashed, bare except for a chair. She was bewildered by two things: there were bars on the one small window, and she was held down in the bed by several straps fitted to the iron cot.

It had taken her a little while to understand that she was confined in the asylum wing of the hospital.

A number of children in St. Pierre this morning had found a new game to play. Soon after breakfast they had massed, about thirty of them, outside the closed door of cobbler Léon Compère-Léandre's shop.

They had begun the game by jeering at Léon, now in his second day of self-immolation. Extending the boundaries and the vicarious limits of the game, the children had started pelting the cobbler's door with stones. Soon a steady fusillade was raining on the door, accompanied by childish screams of excitement.

Suddenly the door opened. Léon, the animal lover, stood there enraged, blinking in the warm sunlight. A stone caught him on the face. With a bellow of rage, he burst forth from the doorway, bent on laying hands on at least one of his tormentors.

The children scattered across the square, fear mingling with their screams of excitement.

Léon spotted one he believed was the ringleader and concentrated on chasing the youth. Together they raced across the Place Bertin, the light-skinned cobbler gaining steadily. The chase entered a side street leading down to the water front. From ahead came the sound of surging water. It was the Roxelane. At this point the river flowed between steep banks of stone. Already the flood water was lapping against the edge of the bank, and in places it had slopped over, turning the ash into a greasy gray mud. As the youth turned to run along the bank of the river, he slipped on the mud and skidded helplessly into the water. He was caught up in the animal and vegetable debris and carried swiftly down-river and out to sea.

The Roxelane, activated by the pressures building up inside Pelée, had claimed its first known human victim. But by midmorning on this sunny, hot Saturday, the citizens of St. Pierre would be stunned to see several other bodies being swept down from the island's hinterland and out into the sea, and only then would some of them begin to feel uneasy about the root of it all: the majestic and, for the moment, silent mountain looming above the town.

In his office in the Governor's Residency in Fort-de-France, Louis Mouttet looked forward to the weekend holiday. All morning, only one report had come from St. Pierre. It was short and to the point, sent by the garrison commander. It said: "The crater is completely inactive."

Later there had also come another laconic report from the commander stating: "Roxelane River in flood, but no immediate danger to St. Pierre." It made no mention of the corpses, dead animals, or uprooted vegetation that the river was swirling out to sea.

There had been one other message sent down the telegraph that morning: it carried the news of the defection of the villagers of Le Prêcheur.

Mouttet had telegraphed back: "Let them go." Their

departure had extricated him from a difficult situation. He could not have held them much longer without causing some comment in St. Pierre; out of town, the villagers could spread alarm among comparatively few people.

Now, as the hands of the clock on the fort which gave the town its name came closer to noon, Mouttet gathered toward him a sheaf of papers. He shuffled them into a neat pile and placed them in a folder; he secured the folder beneath a heavy paperweight made of glass.

Suddenly, "everything on the desk started to dance before my very eyes." For a brief moment the whole room appeared to tremble. Then the shaking ceased, and all was still and normal again—except for the paperweight, which had been cracked by the tremor. Slivers of glass had fallen on the folder beneath. The folder contained the papers on the case of Louis Auguste Ciparis, the man Mouttet had the power to reprieve from the gallows.

Father Alte Roche was on his way to the meeting with Fernand Clerc. He was not optimistic as to its outcome. He felt that the planter could hope to convince few people of any danger from Pelée after this morning's editorial in *Les Colonies*.

The Jesuit could only admire the stand Clerc was planning to make. As the official candidate for the Progressive Party in the north of the island, he was preparing to fly in the face of the Party line and speak out against the potential threat that Pelée offered. In doing so, mused the priest, Clerc was virtually bringing to a close the uneasy relationship he had with *Les Colonies*. An attack on the newspaper would be interpreted as an attack on the Party itself and might even be taken as an open declaration of disagreement with the Governor. Clerc was powerful, but was he powerful enough to survive a head-on collision with the forces which would be clearly lined up against him? Defeat would cost the planter dearly; yet what else could he expect? The priest could find no ready answer to that question.

Father Roche's route took him through the old quarter of the town, through the Bazar du Mobilier and the Bazar Sans Rival, across the Place Crocquet, and into the Rue Victor Hugo. He was halfway to his destination when the tremors rocked the town. They came in two waves, short and sharp, and when they had gone, Father Roche had revised his opinion about the danger of Pelée. He knew enough about volcanic behavior to realize that an earth tremor frequently preceded a major eruption. He, at least, would back any plan that Clerc would offer, providing that it was reasonable and did not carry any overtones of panic, like a mass evacuation of the town. That, reasoned Father Roche as he hurried on his way, would be pointless; it would be far better, he felt, to bring in the populace of the surrounding villages to St. Pierre, where they would be safer than in their homes at the feet of Pelée.

Léon Compère-Léandre was still in a state of shock at the tragedy of the drowned youth when the tremors struck. He was standing on the quayside, looking out to a sea that was covered in a crust of white ash, among which floated the human and animal flotsam that the Roxelane River had disgorged.

The first shock wave knocked him to his knees. A few yards away the second tremor toppled a stack of sugar and rum barrels into the sea.

In other parts of the town, the shock waves produced a milder reaction. In the Place Bertin they sent ash shivering across the square; in the Cathedral of Saint Pierre, they knocked the heavy gold candlesticks off the altar; in the mulatto quarter they cracked several walls; in the press room of *Les Colonies*, they rattled the type in their holders.

Strapped down, Suzette Lavenière screamed in terror as her whole room rocked. Nobody heeded her cries. The medical and nursing staff had their hands too full coping with the sick to spare any attention to a patient whom they believed to be only mentally unbalanced.

Tumbled from Bed

Not far, yet a whole world away, Julie Concoute was also in bed—in a back room of a water-front brothel. The shock waves tumbled her and a sailor out of bed and onto the floor. Picking herself up, the prostitute looked down at her client and roared with laughter.

CHAPTER FIVE

Lava in the Throat

BY MIDDAY EVEN the most optimistic of the villagers had come to accept the conclusion that it would be impossible to reach Le Prêcheur this day. The optimism had started to drain away as soon as they had reached the banks of the Blanche River. The stream they had easily forded the previous day was now a cascade of mud, rocks, boulders, and debris racing down to the sea. Its banks were no longer definable near the coast; at the mouth of the river, the water had spread itself over a quarter of a mile of ground.

Led by their priest but no longer singing, the villagers had started to trek inland. They noticed that though the river narrowed considerably as they went farther from the coast, the flow of the flood water became faster until, a mile inland, it could be seen to be boiling as it came down the slopes of Pelée.

It was then that all optimism vanished. The villagers of Le Prêcheur were cut off from their homes by the torrent. For an hour they sat near the river, debating what to do. They ageed that to return to St. Pierre would be to return to a confinement they could not tolerate, but they could not stay beside the Blanche all day. Finally they decided to make their way inland and across country to Morne Rouge. There they would seek refuge at the Convent of the Order of Notre Dame.

They gathered themselves together and started to move off. Away from the sea, the air was still and sultry.

Lava in the Throat

Everywhere they could see dead vegetation and animals that had been choked by the ash.

The villagers had gone about a mile on their way when they heard a dull roar from up the side of the mountain. They turned and saw an avalanche of mud hurtling down the water course. As it traveled, it folded back on itself, rearing to form a wall of boiling brown in which were imbedded rocks and trees. Impelled by the forces inside Pelée, the moving wall surged its way to the sea.

The villagers were still standing there, mutely disbelieving what their eyes saw, when the tremors which had shaken St. Pierre, three miles away, shuddered through the ground on which they stood, bowled them over, panicked them, and sent men, women, and children scattering across the landscape.

The priest shouted for his flock to return. But, temporarily mad with terror, emotionally and physically exhausted by the past twenty-four hours, the villagers of Le Prêcheur didn't hear him. Many of them ran until they could run no more. Then they dropped into the dust and wept.

There was no one to heed their cries. Several hours before, the people who lived in the country to the north of St. Pierre had also evacuated their homes, leaving their livestock behind to die of thirst or ash fall, and made their way to the town. There they believed they would be safe. The panic-stricken villagers also began to make their way slowly back to St. Pierre.

By midday, the population of the town had increased by several hundreds. The newcomers roamed the streets, blocking traffic, disrupting the commercial life, and creating a bonanza for the café and restuarant owners. Many of the country people told stories of crops being blighted and their houses filling with dust. But in St. Pierre, the stories took on a different meaning than they had in the countryside. Here, the stories were offered as proof that the safest place to be was in the town.

All morning Auguste Ciparis had stood on tiptoe to peer through the barred slit that allowed light into the cell. It was six inches high and twice as wide. Four stubby bars were set in this gap, reducing the light but giving Ciparis a handhold to support himself by as he peered through the grill.

Every morning at dawn, his view had begun with the boots of the long term trusty prisoners scuffing across the courtyard from their cell block on the far side of the jail. Then would come the clanging sound of cell doors being opened as the trusties let the other prisoners out for roll call.

The prison at St. Pierre, like so many French colonial *bagnes*, was run by the prisoners. The administrative personnel were kept to a minimum. The routine discipline in the cell blocks, the running of the kitchens and the hospital, the detailing of the working parties—all these were the responsibility of the trusties. At night when the prison gates were locked, apart from the duty warders in the gatehouse, the *bagne* was entirely a prison society. It was a system that worked well for the penal administration. They were not concerned with any abuses.

There were many abuses in the prison at St. Pierre. Trusties used their positions to promote and further their own ends. Inside the prison they ruled a jungle, a no man's land where bribery and corruption was commonplace and where those who resisted this hegemony suffered violence and even death.

The trusties were either *colons* or mulattos. They had one thing in common, a deep hatred of Negroes, who made up the bulk of the jail's population. It was a hatred which the administration encouraged. In many ways it made the prison easier to control. The trusties would give short shrift to any troublemaker.

Auguste Ciparis had been labeled a troublemaker as soon as he was lodged in the condemned cell. He was a Negro. He had killed a white man with a cutlass. His trial and sentence had caused tension outside the prison. There were rumors that the island's Governor would pardon him to ease that tension. Auguste Ciparis, nineteen years old, coal black and massively built, was definitely a troublemaker by any trusty yardstick.

On his first night in the condemned cell, a party of trusties had opened its door and beaten him with wooden

Lava in the Throat 57

butts. In the days that followed, there had been other beatings. Sometimes it would be a single beating from a warder; usually it was a group beating by trusties.

But in the last two days the beatings had ceased. Ciparis had been left alone. Instead of relief, a feeling of despair had settled over him. For shortly before the beatings had stopped, the prison governor had visited his cell and said it seemed "unlikely" there would be a reprieve. For the past two days Auguste Ciparis had listened in vain for the sound of wood being hammered into the shape of gallows.

This morning he had seen the wood stacked in the far corner of the prison yard as the first soft whites of dawn had begun to lighten the night sky.

Shortly after the discovery, Ciparis had watched the trusties trudging across the yard, their feet sending puffs of lava ash eddying in their wake. Then from above had come the sound of cell doors opening and feet tramping down iron stairs. Soon the prison yard had filled with squad after squad of men, nine deep and six abreast. Ciparis had watched their feet shuffling and then coming to a standstill as the roll call started. By sunup the morning check had ended, and the 269 prisoners went about their tasks in various parts of the jail.

Silence came over the yard. Soon his eyes ached from the glare of the sun reflected off the ash, but Ciparis had gone on watching the pile of wood. For the last two days he had refused food. The fear of dying had killed his appetite.

In the middle of this morning, that fear had reached a new peak. A party of prisoners had appeared by the pile of wood. They had started to assemble the gallows; they worked steadily, seemingly indifferent to the purpose of their handiwork. By midday they had completed the framework of the platform and rigged up the twin posts.

Then a whistle blew, and from all over the prison the inmates made their way to the mess hall for lunch.

The gallows building party had barely left the yard when the earth tremors came. One moment the platform was intact; the next it was a shambles of wood.

To Auguste Ciparis it was a miracle. On the twenty-eighth day of his imprisonment, he had been given a sign that all hope was not lost.

By midafternoon Fernand Clerc had sensed a mounting feeling of unreality around him. Three long hours had passed since the meeting had started. Promptly at twelve-thirty, just twenty-eight minutes after the tremors had swept the island, eight of the most influential people in Martinique had assembled in Clerc's large office on the second floor of 167 Rue Victor Hugo. As each one arrived, Clerc had felt his hopes rising. To him their presence was testimony that the attitude expressed in *Les Colonies* toward the danger of Pelée had been rejected.

"All morning, gentlemen, I have been receiving reports from all quarters around Pelée, and in the town itself. They offer little comfort, and indeed indicate a developing situation that is serious. . . ."

They had listened carefully as Clerc started to review the situation. Saint-Cyr, owner of the Bank of Martinique, sat at the far end of the table, a bluff, portly man, bearded, a gold watch chain draped across his paunch. His presence gave the stamp of authority to the gathering. He controlled one of the financial purse strings of the island; the other string was represented by Emile Le Cure, general manager of the English Colonial Bank, who sat beside Saint-Cyr. Opposite him was Roger Fouché, mayor of St. Pierre and a close friend of Mouttet. Beside Fouché sat the saturnine Alfred Descailles, who virtually had a monopoly on resupplying ships which put into St. Pierre. The group was completed by the Theroset brothers, Pierre and Joseph, casket makers to the island; Georges Madeneux, the town's leading notary; and Father Alte Roche.

Clerc swiftly reviewed the broad situation: twenty bodies had been sighted in the Roxelane and four times that number of dead animals, all of which had been swept down through the town and out to sea; the earth tremors had disrupted the telephone service in St. Pierre; the island was apparently cut off from telegraphic communication with the outside world "by cause unknown"; the streets of St. Pierre were littered with animals and birds, dead or dying from sulphur poisoning, and no amount of clearing away of carcasses by troops could allay the fears of many people that the ash falls could eventually harm them.

It was then that Mayor Fouché interrupted, and the first twinge of unease touched Clerc. The Mayor said that, far from causing harm, he had been informed that "there is medical evidence to show that the sulphur can be beneficial to chest and throat complaints." Endorsement of this remarkable diagnosis came from Joseph Theroset. Since the ash falls had started, he had "personally found that I can breathe easier than I have done before by holding a handkerchief across my nose and mouth to filter the fumes." The fumes, he said, reduced the humidity in the air and could be of no possible danger to any human being. That they had killed animals and birds was simply "because they have a different breathing system than humans."

Clerc continued his presentation. But at the back of his mind the seeds of doubt had firmly taken root: his listeners would not, or could not, comprehend the true situation as he saw it.

"There have been several lava falls. One of my estate managers reports encountering a hunting party descending from Pelée with news of yet another flow. . . ."

All of the twenty-two rivers around St. Pierre, he continued, were in flood; the town's hinterland was being evacuated, and refugees were pouring into the town.

"But gentlemen, is this wise? How do we know, how does anybody know, that we are safe in St. Pierre?" he had concluded.

The question hung in the air for several minutes.

They looked at each other, and then they looked at Clerc. And in their attitude, in their silence, the planter had detected suspicion. Even Saint-Cyr looked uneasy.

"Are you suggesting that St. Pierre itself is in danger?" The question came from the mild Emile Le Cure.

"Yes."

Once more there was silence as each man weighed the answer and its potential effect.

"What will you have us do? Evacuate the town?" Pierre Theroset asked.

"Yes."

This time the flat reply caused a hubbub of excitement and disbelief. To evacuate the town was "impossible," said Mayor Fouché; where would the population go? Fort-de-France? "Unthinkable," said Madeneux. The two bankers said that such a move would cause an economic and social upheaval from which St. Pierre would take a long time to recover. People would "lose faith in those of us they look upon as offering leadership," Saint-Cyr said. "In return we would all suffer from such a move."

Silence returned to the room again as seven of Clerc's visitors reviewed in their own minds the effects a mass evacuation would have on them. It was the eighth who finally spoke. Calmly and precisely Father Alte Roche demolished any argument for emptying the town of its population.

"Today, in view of the tremors, there is likely to be a lava flow," said Father Roche. But the flow, he predicted, would not harm the town. Its likeliest course was down the northern slopes of Pelée, the route previous, prehistoric flows had favored. Even if the lava did flow, as it had "briefly done so," down the southern slopes when it had cut off the Lavenière estate, there could still be no danger to St. Pierre. To reach the town, the flow would have to cross three valleys, each of which was more than one hundred feet deep. If the flow came down the western slopes, immediately above the town, it would again be checked by the contours of the land.

"Therefore, if there is to be an evacuation, it must be an evacuation of the people in the outlying areas, who should be brought into the safety of St. Pierre," he concluded.

After some discussion, Clerc's visitors decided on an immediate course of action: they would form an "action committee." Saint-Cyr would be its president, the rest would be members. Their task would be to supervise the settling-in of all those who sought safety in St. Pierre.

Now, late in the afternoon, his visitors departed, Fernand Clerc stood alone in his office. He felt exhausted and frustrated. The meeting had been a crucial opportunity to alert the civic leaders to the threat he believed was contained in Pelée's behavior. He had failed. His sense of defeat made him feel almost physically sick.

A mile away in a narrow building which housed the headquarters of the Radicals, Senator Amédee Knight was briefing the party workers. The room was jammed with men and women of various shades of black, with an occasional white face. They listened respectfully as the Senator talked; to them, as to all the colored population on the island, Amédee Knight was the black boy from the plantation who had shown the world that brain power was no special property of the whites. The fact that Knight *owned* rather than *worked* on a large plantation in the southern end of the island and was reputed to be a far harder taskmaster than most white employers made little difference; he was a black man who had resoundingly beaten the whites at their favorite pastime, politics.

The first signs of life in the volcano had been observed on the opening day of the campaign. To Amédee Knight, plotting and scheming for a Radical victory despite a split among the party's supporters, Pelée offered new hope; for him "the smoke of Pelée became a victory sign. From the beginning I saw that continued life in the volcano would bring in the Radical voters from the countryside to the safety of St. Pierre. In this way the supporters of both candidates would be in one camp, and the physical split in our ranks would be healed. There was next the question of making all sides believe that it would be best after all to support one candidate. The Party hierarchy had made its choice, Percin. It was now only a matter of persuading Lagrosillère's supporters to unite in the common aim of victory."

Senator Knight, ever ready to seize an opportunity, again showed his generalship, as he "instructed Lagrosillère that while he was still to stand, simply because it was too late to withdraw his nomination, Lagrosillère was to agree that in the event of an inconclusive result, he would not compete in the second ballot, but would ask that all who had voted for him should then support Percin." In return, Lagrosillère was assured of high office when the Radicals came to power, as Knight was certain they would.

They had been helped by several things. For a start Fernand Clerc had given a surprisingly poor impression in the campaign. His command of oratory seemed to desert him before a mass audience; his speeches bore the marks of uncertainty and rewriting. Audiences began to dwindle as he deviated further and further from the Progressive Party ticket. The virulence of *Les Colonies* had helped to unite the Radicals; frequently the viciousness had boomeranged, hardening the resolve of the colored voters to drive white bigotry from office. For the most part the Radicals relied on slogans and posters to put across their arguments, effective methods to an electorate that was largely illiterate. For the past three weeks the streets of St. Pierre had echoed with a score of catch phrases.

All these factors had helped the Radical cause. But the real assistance had been Pelée.

To a black population rooted in superstition, the volcano's awakening was seen as a "sign" of a need for change. Amédee Knight had given orders to play up Pelée's influence. Another slogan had been born: "The mountain will only sleep when the whites are out of office." Among the popular reasons given for the awakening of the volcano were that it had stirred in protest against the rapacious white business houses, that it was a warning to the whites to show "more Christian tolerance" to their black brothers.

Here, as in so many issues, the electoral differences dividing Fernand Clerc and Radical policy were blurred; both were often making an appeal to the voters to end St. Pierre's role as the Sodom of the West Indies. On better education and working hours for nonwhites, Clerc and the Radicals were close enough for Amédee Knight to have remarked: "If only he was black he would be a Radical!"

The strategy that Amédee Knight outlined reflected the confidence he felt: every Radical supporter among the 6,164 voters on the electoral role in the northern *arrondissement* was to be taken to vote. The primary would be simply a "dry run" for the real test a week later, when the final ballot would be held between Clerc and Percin.

Like Fernand Clerc, Senator Knight was coming to the conclusion that Pelée would play an increasingly vital part in the events to come. News of the planter's failure to enlist support to evacuate the town had been brought to Senator Knight shortly after the "action committee" had been formed.

"The political gain to be made out of Clerc's failure was instantly obvious," Senator Knight was to admit later.

Clerc's sincere call for evacuation would, in the days to come, be deliberately misconstrued. The first shot in the distortion was fabricated late this Saturday afternoon in the Radical Party headquarters when yet another slogan was devised:

"Pelée demands all whites leave St. Pierre—or death for us all."

In the last light before the curtain of tropical darkness dropped, Thomas Prentiss conducted the most unusual conversation he had ever had. The American diplomat stood on the bank of the Roxelane River opposite the British Residency. About twenty people were grouped around him, listening to his dialogue with the British Consul General, James Japp, who stood on the outer wall of the Residency. Japp was having difficulty in hearing. The flood water surging against the wall was a roaring torrent between them, drowning most of what Prentiss said.

Tired of shouting against the noise, the American turned to a mulatto and thrust a broadsheet into the man's hand.

"Can you read?" demanded Prentiss.

The man nodded.

"Then read this in your loudest voice to M'sieu Japp," commanded the diplomat.

The man studied the paper with care. Then, satisfied, he started to read:

"Les Colonies. EXTRAORDINARY EDITION."

The man paused to let the full import of the words convey themselves to his audience. The word "edition" was a grandiose one for the single broadsheet that bore all the hallmarks of haste.

Ten thousand copies of the extra had been printed

late that afternoon. They had been done on the instigation of Robert Saint-Cyr and his "action committee." It was their content that had brought Thomas Prentiss hurrying to the river bank, eager to discuss it with his marooned fellow diplomat.

The mulatto was a slow reader, but his voice was strong, carrying above the flood water:

MOUNT PELÉE AND ST. PIERRE. Yesterday the people of St. Pierre were treated to a grandiose spectacle in the majesty of the smoking volcano. While at St. Pierre the admirers of the beautiful could not take their eyes from the smoke of the volcano and the ensuing falls of cinder, timid people were committing their souls to God. . . .

The mulatto paused, peering around him, seeking assurance that his words were completely understood. Prentiss waved for him to continue.

It would seem that many signs ought really to have warned us that Mount Pelée was in a state of serious eruption. There have been slight earthquake shocks this noon. The rivers are in overflow. The need now is for the people outside St. Pierre to seek the shelter of the town. Citizens of St. Pierre! It is your duty to give these people succor and comfort.

Again the mulatto paused, and the group gathered on the bank waited expectantly. The man resumed reading:

Because of the situation in the hinterland, the excursion to Mount Pelée which had been organized for tomorrow morning will not leave St. Pierre, the crater being absolutely inaccessible. Those who were to have joined the party will be notified later when it will be found practical to carry out the original plan.

The mulatto handed the paper back to Prentiss. Nobody said anything. Then, for the second time that day, the American diplomat said: "I think the world has gone mad."

Japp did not hear him. The Englishman had already lowered himself behind the Residency wall. As he did so, the last of the light went.

From the bridge of the *Topaz*, its worried master, Sequin, watched night come to Fort-de-France. He had just left the Governor's Residency. Mouttet had been waiting for him in his study. Even with the late afternoon sun, the day had been gloomy as Sequin was shown in.

Mouttet was flopped in a deep chair, listless, and the sailor realized that a change had come over the man who had sent him off to St. Lucia the day before. Mouttet looked older, grosser. But it was his voice that shocked Sequin. The raucousness and harshness had increased.

"Well?" he had demanded.

Mouttet had almost appeared not to be listening as Sequin reported that the telegraph office in St. Lucia had informed him that the cable link between Dominica and Martinique had snapped that afternoon. The seismograph in the harbor superintendent's office had recorded a "quite severe tremor in the area of Martinique."

"Is that all?"

Sequin nodded, anxious to be on his way, frightened by the Governor's demeanor.

Then Mouttet began to speak.

"There have been other tremors today. They did not harm anybody. There is nothing to worry about. Yet everybody is in a state of unease. In St. Pierre they have talked of emptying the town!"

As Mouttet spoke, his voice rose. Panic, he said, was being whipped up for political reasons, but it would do no good. Soon the volcano would subside and life would return to normal.

"More, my dear Sequin, is being made of Mount Pelée than it deserves," he shouted, pulling himself from his chair. "There is nothing to fear!"

Mouttet started to walk slowly round the room. Perspiration beaded his face. His body, Sequin saw, shook under the strain of nervous excitement.

"I have run this island the way people have wanted it

run. I will go on running it. And nobody, nor anything, will stop me!"

Exhausted by his outburst, the Governor slumped back into his chair.

For a while Sequin had stood there, not daring to speak. Then at last he realized the interview was over. Mouttet had closed his eyes and appeared to be asleep. Quietly, Sequin had left the study.

Now, standing on the bridge of his ship, the sailor could think of only one explanation for the Governor's behavior. Louis Mouttet had lost touch with reality.

In the splendid isolation of their hillside home, Fernand Clerc was confiding the same diagnosis to his wife, Véronique.

They sat alone at opposite ends of the long dinner table. Between them were bowls and dishes of gumbo, crab claws, pompanos, yams, okra, and red snappers.

Normally Clerc had a good appetite. Tonight, reviewing the events of the day to his wife, he had eaten nothing.

He recognized the extraordinary edition of *Les Colonies* for what it was: another example of the form of political byplay that the newspaper thrived on. In recognizing it, Fernand Clerc also admitted that he had no stomach for the crude cut-and-thrust that is an essential part of any politician's armory. His whole background, with its concentration on business and accepted social position, placed him in a personal position far removed from the Progressive cohort.

Cautious and devoid of flamboyance, he was a man ready to probe general principles in depth. In doing so he had inevitably challenged many of the fundamentals of Progressive policy. To him many of their ideas were not intellectually respectable and could not be absorbed into the context of his own political ideals. He recognized, too, that he could count on little personal support, because his political ideas appeared remote to the man in the street— who relied upon gossip, prejudice, and shibboleth served up by *Les Colonies*.

Yet he was determined to go on to the bitter end. He was determined to convince his fellow citizens of the danger they faced by remaining within erupting distance of Mount Pelée.

Abruptly, he rose from the table and made his way to their bedroom balcony. He stopped before the barometer fixed to the pillar. By the light of a match he studied the needle. It was trembling. He turned and looked across to the volcano, softly backlit now by the rising moon. Then he gazed down at St. Pierre; at that moment—"almost as if I had a vision" he was to remember—he knew the town was doomed.

He was still standing there when his wife joined him.

"Tomorrow, we will go to the Cathedral and pray for deliverance," she said.

"It may be too late," he replied.

"All the same we will pray," she insisted, "and for victory in the election, too."

In the mulatto quarter prayers were being offered for the safekeeping of the soul of Eli Victor. He had died that afternoon, after a brief illness, in the military hospital. He was as poor in death as he had been in life. He had walked barefoot all his days. Now there was no hearse, no car, no cart, not even a mule to carry his body home.

Instead he was leaving the hospital on a crude stretcher made from sacking and two bamboo poles, carried by four men. They carried him past the barred window behind which Suzette Lavenière lay strapped to her bed. The bearers walked slowly, as befitted a ritual dedicated to appeasing the departed soul of Eli Victor so that it would not haunt the living.

It was a ritual known as the "nine nights." Its roots lay in ancestor worship and voodoo. Among the upper classes it had degenerated into something like a wake. The nine days and nights after death were used to bring succor to the bereaved relatives. But to the smallholders, the barefoot people in the mulatto quarter, the ritual meant more; in the case of Eli Victor it meant giving his soul a proper entry into Heaven.

In the darkness the pallbearers set off, singing softly among themselves.

For Léon Compère-Léandre the day had ended as it had begun, in the confined isolation of his cobbler's shop in the Place Bertin. He had returned there late in the afternoon, shocked by the events of the day and determined to stay in his shop until the volcano subsided.

The shop sloped downward from the Place Bertin, until at the rear where Léon squatted amongst his lasts, hammers, and boxes of nails, it was several feet below the level of the cobble-stoned square. He was shut off from the outside world by a wooden door. It was made of ship's planking. A year earlier Léon had reinforced the outside of the door with several layers of tin sheeting as a protection against the wood's rotting in the humid air. Now, it was a piece of handiwork which he hoped would stand guard against Pelée.

This Saturday night, the door blocked out the sound of the funeral procession singing as it made its way across the square to the mulatto quarter.

In the Orphanage of St. Anne, twenty-eight children slept peacefully. Earlier that afternoon they had been told that school lessons had been suspended until the volcano had settled down. In the morning they would be part of the choir in the Cathedral of Saint Pierre, as children from the orphanage had been for the past fifty years. Tomorrow would be the last Sunday their shrill voices would be heard.

On the far side of the Pont Basin, an iron-girdered bridge over the Roxelane River that led into the mulatto quarter, a sizable crowd had gathered in the darkness. They were the main mourning party for Eli Victor. At last they heard what they had been waiting for: a far-off murmur that began to grow. In the space of a dozen breaths, the waiting crowd could make out the words of the funeral chant. In another moment the keening harmony was lapping at their ears. Someone struck a match and naked flames flared, the torches casting long shadows as they spluttered to life.

A humming rose from the group and became singing

Lava in the Throat 69

in answer to the approaching singers and in welcome for the dead.

At last the pallbearing party came in sight, rounding a bend in the road. The singing of the crowd rose and deepened. Eli Victor was being brought to his grave, held high by the bearers.

The crowd crossed the bridge to merge with the approaching party. Fresh shoulders eagerly took up the burden of Eli Victor's body. Then, singing as they went, the procession started to cross the bridge. Bare feet trod through the ash in soundless rhythm, and Eli Victor rode in their midst like a Pharoah.

The pallbearers were halfway across the bridge when the tremors came. One moment the bridge was a solid structure; a splinter of time later it had been wrenched from its supports by a twitch of nature. Twenty men, women, and children, including the widow of Eli Victor, were pitched into the raging water below, to be swept out to sea with the corpse they had come to honor.

On the bank, the tremors tumbled other mourners into heaps. As they picked themselves up, they faced a new terror. The sky over Pelée's summit was starting to turn a bright red.

A mile away, from the window of his Presbytery, Father Alte Roche watched the reddening sky. He knew that the glow was a reflection of the lava rising up inside the throat of the crater, turning the rock around the rim a dull red, crowning the crater with a fiery cap. Soon a sliver of that crown started to detach itself and move slowly away from the rim. Its progress down the side of the volcano heightened the feeling of peculiar satisfaction which occupied the priest. The lava was moving down the barren northern slope of Mount Pelée, away from St. Pierre, along the route prehistoric flows had favored. His prediction had been accurate.

He turned from the window and sat at a wooden table. On the table was a small book. It contained the Jesuit's observations of the volcano's behavior since the eruptions started. Now, under the heading "Saturday, May 3, evening," he wrote: "The nature of the eruption has changed. The mountain has now vomited its first real flow of lava. Frightening though the glow appears, it presents no real threat. So far there is no noise, and the whole impression is one of beauty."

Having brought the log up to date, Father Roche placed it inside a steel trunk underneath his bed.

Outside, the glow from Pelée's throat grew richer. It was indeed a thing of beauty.

CHAPTER SIX

The Whole World in Flood

THE SEISMIC ACTIVITY which had sent the mourners tumbling and wrecked the Pont Basin had confined itself to the downtown seaboard of the mulatto quarter. There it had caused considerable fear. Clocks had stopped their pendulums, tables had tilted, and doors had opened. Crockery here and there fell from shelving, and ceilings swayed as if suspended in free air.

Julie Concoute, the seventeen-year-old prostitute, was making her way down from her room to solicit a fresh client in the Bar Normandie which occupied the entire ground floor of the water-front bordello where she worked when the tremors pitched her down the wooden staircase to the floor of the bar. She lay there for some time before she was discovered and carried to a room at the back of the bar. Her neck was broken. She was dead.

By the time the panic had passed, eight other people had been killed by the tremors. Five of them had been struck by pieces of falling masonry. The other three, all children, were crushed on the water-front beneath a stack of rum barrels which had been toppled over by the tremors.

To the south of Martinique, in the Meteorological Observatory on the island of Trinidad, the duty officer, Edward Paice, recorded the tremors as having a "markedly horizontal concussion, but it is impossible to distinctly locate the quarter of first impact except that the oscillations are registered as north-west to south-east."

It was a routine observation. The records of the Observatory were filled with such observations. It would be nearly five days before Edward Paice would realize that his observation had been the first scientific recording of the disaster of St. Pierre.

St. Pierre this Saturday night had taken on a bizarre quality. News of the effects of the tremors had traveled to several parts of the town. Instead of spreading panic, the news provoked a strange calm. It was one of the curious inexplicable side effects of a day of drama. People started to make their way to the mulatto quarter, many undoubtedly attracted by the fascination that tragedy always had: they wanted firsthand accounts of what had happened to the mourners for Eli Victor, and how Julie Concoute had died.

To Philomène Gerbault, a cousin of the island's Deputy Governor, the night had assumed a particularly personal unreality.

For the past half hour she had been trying to reach the Cathedral of Saint Pierre. Three times she had almost reached its entrance, only to be caught up in a surge of people coming the other way, forcing her back. Now, this small, wiry woman was making one last effort to reach the Cathedral.

The journey was getting increasingly difficult: the streets were filled with an influx of country people who milled around aimlessly, seeking shelter or reassurance. Many of them, including the villagers of Le Prêcheur, had made their way to the precincts of the Cathedral. By nightfall the area around the Cathedral was a seething mass. Already troops from the garrison had started to clear the area.

Finally, halfway down the Rue du Collage, she admitted defeat. She turned and joined the tide of men, women, and children and was swept back the way she had come. At the end of the Rue du Collage, she allowed herself to be guided into the comparative still water of the

Rue Bouille. Halfway down the cobbled street she had an apartment. Tonight Philomène Gerbault came to a decision that others were to endorse in the next few days: St. Pierre was no longer a safe place.

But for every person who left the town, three others would come in from the countryside around. The population of St. Pierre had swollen to nearly thirty thousand. Even under normal circumstances the town would have found it difficult to cope with such a sudden increase; now, with food running scarce, drinking water polluted by ash falls, and livestock dying in the streets, the situation was steadily deteriorating.

Philomène Gerbault, forty-seven years old and a widow, wanted no part of this. Arriving at her home, she ordered her maid to pack a trunk and get the stable men to prepare her carriage. Thirty minutes later, seated in the back of the carriage with her mail, she set off for her other home in Fort-de-France.

Later, she was to recall that "the passing of the carriage made no noise. The wheels were well muffled by the ash. The horses, removed from the shelter of their stable, snorted from the effects of the sulphur fumes. The streets were hard to negotiate. People would not give way, and it needed a lot of patience and firmness to force a passage."

The journey out of St. Pierre was an unforgettable one for the widow. She was to describe it later to Angelo Heilprin, a Fellow of the Royal Geographical Society of London. "One can hardly imagine a more hopeless scene of impending ruin; for what the volcano had thus far spared, or seemed disposed to spare, the torrential waters of the descending streams threatened to claim. Birds lay asphyxiated by the ash. The cattle suffered greatly. Children wandered aimlessly about the streets with their little donkeys, like little human wrecks. We passed a group of children going hesitantly down the Rue Victor Hugo. They looked as if they were covered with hoar frost. In the countryside, desolation and aridity prevailed. Little birds lay dead under the bushes, and in the meadows the few living animals were restless—bleating, neighing, and bellowing in despair. But as we traveled deeper into the countryside, an eternal silence seemed to envelop everything. It was all the eerier when viewed by the light of the glowing cone of Pelée."

The carriage was a quarter of the way to Fort-de-France when the silence was broken by a series of explosions: "They were frightful detonations. And then we observed one of the most extraordinary sights in nature—Mount Pelée awake at night. The glowing cone was soon hidden by an enormous column of black smoke traversed by flashes of lightning. The rumbling grew deeper. A few moments later a rain of ashes fell upon the countryside."

Without waiting to see what happened next, Philomène Gerbault, her maid huddled in fear beside her, ordered the carriage driver to make all haste to Fort-de-France.

Behind her in St. Pierre anarchy was beginning to corrode the mechanism which kept the town functioning. In the Mouillage and Fort districts of the town, several food and vegetable shops were looted; soldiers were called to evict people from a number of hotels and bars where they had gone demanding free food and shelter; fights broke out between refugees from the countryside and local residents who refused to open up their homes as sanctuaries.

The arrival of a black cloud of warm ash put an end to the looting and fighting. The ash fall had the density of a thick fog, and the effect of a cloud of tear gas. Within moments, St. Pierre was choking in tears.

In the darkness the roar of the water as it swept past the Residency sounded to James Japp, the British Consul-General, as if the "whole world was in flood."

It would be some hours before he would learn that a miniature tidal wave had coursed down the Roxelane River, wrecking as it passed the power plant half a mile outside St. Pierre which provided the town with much of its electricity supply.

In the town, the red-eyed populace had to cope with a new phenomenon. Through the falling ash they could hear the muffled noise of what sounded like heavy artillery fire. The resemblance to gunfire was heightened by the regular flashes that could be detected through the cinder cloud. Mount Pelée was coughing out red-hot rocks that weighed up to half a ton; they arced through the air, shattering down on the countryside below, setting fire to jungle, brush, and deserted hamlets.

Soon the rain of ash was met by a rising pall of smoke; together, the cloud of ash and smoke drifted slowly over the town.

It took an hour to pass. When it had gone, it was discovered that the cloud had choked thirty people to death.

Its passing marked a return of that same eerie feeling of security. A year later, trying to analyse that feeling, Angelo Heilprin was to write: "In any country but Martinique the symptoms of uneasiness to which Mount Pelée gave expression would have impressively counseled flight. But in this island of tropical dreams and sunshine, the warnings were largely ignored."

Now, as midnight approached, people from the villages to the south of the town—from such romantically named places as Carbet, Morne-aux-Boeufs, and Gele— began to move into the town, convinced that within its confines they would find the security they sought against the elements.

From his balcony Andréus Hurard, the editor of *Les Colonies*, watched a fresh group of refugees entering the town. Their arrival gave him a feeling of deep satisfaction, confirming again the conviction that not only had he extricated himself from a difficult situation, but that he had been able to turn it to his own advantage.

A town choked with sulphur, its streets blocked with fallen ash, a burning and thundering volcano standing on its threshold—this was St. Pierre as Saturday drew to a close. This was the town that Andréus Hurard, for a mixture of commercial greed and political chicanery, was determined to hold together until after the impending election.

As he turned to go indoors, the striking of the Town Hall clock announced that it was midnight. Sunday was a few strokes away. In just eight hours, the polling would begin. Hurard had no doubt that at the end of it, Fernand Clerc and Alfred Percin would be the two candidates to go forward to the final ballot a week later. But, like many other people in St. Pierre, he had miscalculated the amount of time left.

Four thousand miles away in Paris, at the *aperitif* hour of seven o'clock, Pierre Louis Decrais, the Minister of the Colonies, sipped a drink as he studied a summary of the week's reports from all corners of France's empire. Born in Bordeaux in December, 1838, he had matured, like many of the wines in the district, early in life. At thirty-three he had been appointed Prefect of Indre-et-Loire; three years later he had become Prefect for the important Alpes-Maritimes; at the age of forty-nine he had become Minister of the Colonies.

He read fast, scanning each page and initialing it at the bottom righthand corner, signifying that it had received Ministerial scrutiny. Occasionally, he penciled in some comment.

Under Martinique was a brief entry: "The election is proceeding. It is too early yet to say how the issue will be decided."

Decrais initialed the page and turned to another report.

SUNDAY

✌

May 4, 1902

CHAPTER SEVEN

Special Powers

AT TWELVE-THIRTY on this Sunday morning, only one light burned in St. Pierre. It was the oil lamp in the parlor of the American Consul's home.

For hours Thomas and Clara Prentiss had sat listening to the sounds of eruption coming from Mount Pelée. At one point Thomas Prentiss asked his wife if she would not reconsider going away from the town, if only to stay in Fort-de-France, until the volcano quieted again. She had firmly refused to leave his side. Since then they had said very little to each other.

Life for the Prentisses in St. Pierre was in many ways an unreal one. Though they were important members of the local diplomatic corps, with all the undertones of solidarity that the title implied, there was an ironic ring about the word "corps" when applied to the foreign diplomats posted in Martinique. There was no official Legation quarter, though most lived in the L'Centre district, and there was little social mingling between diplomats and the local population. Prentiss, like his fellow envoys, was insulated from all but the most formal contacts with the populace; only James Japp, the British Consul, maintained any real social liaison with the local administration by virtue of his length of service.

Apart from entertaining the master of an American ship in port, or inviting Japp to dinner, the Prentisses rarely entertained. This Saturday night they had dined alone. Now in the small hours of Sunday morning, Clara

Prentiss rose to prepare the *café brûlot* she had served as a nightcap since arriving in St. Pierre.

She had started to ladle the liquid into two large silver tankards when a loud explosion echoed across the countryside. Moments later a flash of light glowed brightly through the shutters.

Clara and Thomas Prentiss, clutching their tankards in their hands, opened the shutters. An extraordinary and terrifying spectacle greeted them.

As far as their eyes could see, the whole sky, clear now of the last of the dust cloud, was "aglow with fire."

Looking out through the window, Clara Prentiss felt "as if we were standing on the brink of Hell. Every few moments electric flames of blinding intensity were traversing the recesses of black and purple clouds and casting a lurid pallor over the darkness that shrouded the world. Scintillating stars burst forth like crackling fireworks, and serpent lines wound themselves in and out like traveling wave crests. It was indeed an extraordinary spectacle."

There was another extraordinary phenomenon nearer at hand. It was the attitude of the majority of the population of St. Pierre. The display of celestial fireworks in the clear sky, far from causing panic, only generated excitement. People leaned out of bedroom windows and stood in the streets, shouting and whooping as the sky was patterned various shades of red.

To Robert de Saint-Cyr, awakened from sleep by the noise, the spectacle was reminiscent of the Mardi Gras of his childhood in New Orleans. He stood for a while and watched the macabre carnival mood which had settled over the town, then closed the shutters and went back to bed. Down in the streets, children, now fully awake, raced around, sending swirls of ash eddying into the air.

From his bedroom window, the indefatigable Father Roche carefully recorded what he saw: "The number of forms in which the illuminations appeared were many. Some were short, straight rodlike lines; others were wavy or spirals. There were large stars, and circles with clusters of stars attached to them. And around these zigzagged the lightning. After that first loud explosion, there were no intermittent explosions, but instead a steady roaring that swept around the peak of Pelée, rising and falling almost as if keeping in time with the auroral display, though it is likely that they are the flashes of burning gases escaping from the crater of Mount Pelée."

Like the lava flow of the previous night, the priest concluded, the flashes "present no danger, and appear to be identical with those which were noted to accompany the great eruption of Tarawera, in New Zealand, in 1886."

He was still watching this scene of majestic grandeur when a "great pattering of pumice fell upon the town, and for a while it sounded as if we were in a tropical hailstorm. Some of the fragments were an inch or more in size; most were like fine sand grains. But even the smaller particles came down with great force, stinging the flesh as it was touched by them."

The fall lasted about ten minutes. In that time it had driven all those in the streets of St. Pierre to shelter. When the fall ended, so did the celestial display, and the night returned to normal.

As he dressed on this Sunday morning Dr. Eugène Guérin saw that the wind had veered; smoke from the chimney stack of his sugar factory was drifting lazily northward. The Blanche River, which the day before had carried a wave of mud down to the sea, was now flowing past the factory wall at a rate no greater than when it was in normal flood; the mud had slopped over the wall in places and congealed to form crude pancakes.

Though he had observed it from childhood, Dr. Guérin had never failed to be impressed by the scene of incomparable beauty that greeted him from the balcony of his home.

To the left and above his factory, its lofty peak now veiled in cloud, rose the wall of Pelée. To his right the hills came down to the sea; a ribbon of curving white surf divided sea and land.

Between the volcano and the sea were massed golden-green regiments of sugar cane, rising to twice the height of a man. Dr. Guérin looked out upon what seemed to be endless square miles of stalk, planted so closely together that their tops formed a seamless carpet. Now, under the repeated ash falls, the pile of that carpet was a dull white.

The dust had settled on the sugar processing factory close to the villa. The storage sheds, great open barns roofed with corrugated iron; the loading platforms, where workmen armed with long-handled rakes guided the cane onto a conveyor belt; the processing building; the enormous storage vats—all were thickly coated with the refuse from Pelée.

The fallout had brought its own special problems to Dr. Guérin. Not only had it ruined part of the cane; it had also produced a reluctance among the pickers to harvest the crop on the slopes of Pelée. Twice in the past few days, during ash falls, they had refused to leave their stonewalled homes.

During the early hours of this Sunday, Dr. Guérin and his son Eugène had been called from their beds to reassure the employees that the detonations they heard, along with the lightning and other celestial trickery, was not a prelude to a major eruption. Unlike their sophisticated cousins in St. Pierre, the plantation workers were simple God-fearing people whose lives were ruled by the threat of Divine wrath and damnation. It was a threat that Dr. Guérin, like his ancestors, fostered; fear of the Almighty was a good way to keep a man at his work. Dr. Guérin knew it was this fear, coupled with the very real fear his own presence instilled in them, that kept his workers from fleeing.

But now, in the light of daybreak, he saw that their fears had not been altogether groundless. The lower slopes of Pelée were dotted with fresh craters caused by the massive rocks which the volcano had ejected; where the rocks had shattered, the fragments had mown swaths through the cane.

Around these paths he could now see bits of bright color moving. They were the canecutters, men and women, and their garments—spots of bright yellow, red, purple, and blue—glowed in the light.

Sunday was normally a day of rest at the refinery and in the surrounding fields, as it was everywhere else on

Martinique. But for Dr. Guérin this was no ordinary Sunday. The previous evening he had heard that the captains of a number of cargo ships in St. Pierre were planning to sail with their holds unfilled if the volcano continued to threaten the countryside. If they did, it would mean that his factory's output would have to wait another month before fresh ships called, and a month could mean the difference between profit and loss on the sugar crop. Unless reaped quickly, the cane would be ruined if further ash falls came; there was even the possibility that a really heavy fall could clog up the processing plant. So Dr. Guérin had ordered his foreman to have all hands at work this Sunday if the weather, and Pelée, permitted it. Those who refused were to be summarily dismissed.

Dismissal from the Guérin payroll meant not only a loss of free living accommodation for the sacked man and his family, but also a virtual impossibility of employment on any other plantation in Martinique. For Dr. Guérin at seventy-two was the titular head of the oldest ruling dynasty on the island. He was a member of the Ten Families.

The Ten, and only ten, individual families were the island's social elite. They had their roots in France, tracing their ancestry back to Agincourt when a Guérin, an Aubrey, a Janne, a Hayot, or a Cottrell had distinguished himself. Fernand Clerc, for all his power, was not a member. The Ten Families had two things in common: they never openly became involved in the everyday politics of the island—attempts by the Radical Party to agitate among their workers had been swiftly put down—and they were very rich. The poorest of them had a fortune of around $12,000,000; the richest could command eight times that figure. Their money, like everything else they had, was cautiously handled, invested in the safety of the banks of Paris. They lived, and ruled, by decree.

Nowhere was that autocracy more clearly seen than in the attitude of Dr. Guérin and his family. They brooked no argument or interference from anyone; their two thousand acres of land were virtually a sovereign state.

The center of this state was Dr. Guérin's home, a Caribbean equivalent of a provincial French château. On two sides it was surrounded by gardens filled with frangipani, orchids, and bougainvillaea. Beside the chateau were the house servants' quarters; beside them, the factory; behind the factory, his son's home.

In six days' time, Dr. Guérin would be free of the ash. On Friday, May 9th, he and his family planned to collect his son Eugène's children, who were staying in Fort-de-France, and board the Quebec liner, S.S. *Roraima*, for their annual pilgrimage to France. In all there would be fifty members of the family entourage, including nannies and servants, embarking on the voyage. Eight weeks after leaving St. Pierre, they would be in Paris. There they would go to the same hotel, near the *Gare de l'Est*, where generations of Guérins had always stayed. Throughout their stay, they would consort with none but other socially acceptable families from Martinique or other parts of the West Indies. Their Parisian holiday completed, they would go to Vichy for "the cure," and to Vittel for "the waters." Then they would set sail for home again, having bathed their souls as well as their bodies in France. In many ways Dr. Guérin was as removed from the realities of everyday life as were any of the foreign diplomats on the island. He and his family wanted for nothing.

But in just twenty-four hours' time, his wife Josephine, his elder son Eugène and his wife, his youngest son Joseph, his daughter Sarah, his English nurse Mary Goodchild from Lowestoft, two housemaids Marie and Cece from Dijon, the family cook Georgette—all would be dead. And with them would die 148 others who depended on the sugar refinery for their living.

A mile away death had already struck the village of Ajoupa-Bouillon, which lay at the eastern foot of Pelée. Its three hundred families lived in wooden houses, and almost alone among the country folk they had refused to evacuate their homes.

The village mayor, Bertrand Kloss, who held office by virtue of owning a large cacao estate which employed all the men and boys of the village, had pointed out that far from causing damage, the ash falls from Pelée would give added fertility to the soil.

During the early hours the villagers had gathered to watch the illuminations from Pelée; in awe they had seen the red-hot boulders cascading from the crater to shatter, it seemed to them, close to St. Pierre. None of the boulders had landed near their village, and they had been right to remain in Ajoupa-Bouillon.

They had gone back to their interrupted sleep. A few hours later, shortly after dawn, a number of chipboard houses suddenly collapsed.

The zone of destruction began a short distance below the village church and extended for a quarter of a mile down the slope of Pelée.

It was caused by the ground suddenly cracking open under the immense pressures building up inside Pelée. Like so many other volcanoes in the Antilles, Pelée had been producing for the past week what geologists would immediately recognize as seismic jars. These are really small, localized earth tremors of varying intensity. Pelée, like all the other volcanoes in the Caribbean chain, had a thin crust of rock on its outside which could not resist the deep-rooted strains rising from several thousand feet below. There, the rocks were ground together by the pressure, liquifying in temperatures of over 4000°F. This white-hot mass had steadily forced its way to the surface; in most cases it had been able to find its outlet through the crater, but once in a while the pressures had been too demanding. A jagged crack had ripped open in the ground, releasing a mixture of steam and boiling mud.

The blast from deep inside the earth destroyed all it touched: in a matter of seconds lush vegetation had withered into arid wasteland. Cattle and horses caught in the mainstream of the explosion had been burned to death; others, still alive, had the hides burnt off them, leaving raw flesh exposed to the morning air.

The fissure had burst through a number of the village houses, destroying their wooden frames and instantly killing their occupants.

Those who survived the instant impact of the blast suffered an even worse fate. The steam and mud, ejected with the velocity of a high-pressure hose, had inflicted horrible injuries. In one house a woman rolled in agony in a corner of a bedroom, her flesh burned and "hanging in places from the bone." The steam had gouged out both her eyes, and the boiling mud had skinned a leg down to the bone.

For Bertrand Kloss, leading the rescue operations, she was the first of many of his villagers he was to watch die.

Later he was to remember this Sunday morning as "being too shattering to comprehend."

The first houses that he led the rescuers to had simply exploded off the face of the slope; even their foundations had disappeared.

Further down the village they came to a small cottage. It appeared untouched by the blast. But from within came a deep moaning. The occupants had been cruelly burnt by a jet of steam which had roared up through the floor. To Kloss they called out for water to relieve their raw throats. None was readily at hand. They died as they were gently lifted out into the open.

The rescuers entered another cottage: "A dim taper illumined a nearly black interior sufficiently to permit us to see a writhing figure being tended by the hand of one who was probably dearest to it. The cries of pain were heart-rending. Flies were swarming everywhere about the place, and the odor from singed flesh was almost unbearable."

House after house produced the same picture of horror. In many, "the occupant's flesh was as if baked and steamed, with raw masses appearing where there was no longer skin."

The fissure had killed 158 of the population of Ajoupa-Bouillon instantly. Another thirty of the seriously injured were to die from their burns as they were carried down the slope of Pelée on makeshift stretchers.

Four miles away, in the Members' Lounge of the St. Pierre Sporting Club, the "action committee" was in session. One topic dominated all else: the shortage of food and accommodations in the town to cope with the influx of refugees.

A third of the food, vegetable, and butcher shops in

St. Pierre had not taken down their shutters this morning because they had nothing to sell.

"Others have only supplies enough to last a day at the most," reported Roger Fouché, mayor of St. Pierre, "the *masions* Saint-Eves, Deplanche, Doileret, and Reynoit all announced that they will be closing their doors as of tomorrow."

He had named the four wholesale merchants who supplied the town's shopkeepers with provisions.

"Food must be brought in from Fort-de-France," said Emile Le Cure, the general manager of the English Colonial Bank.

"It will take time to organize," predicted Alfred Descailles, the island's leading produce supplier.

"But it can be done," said Pierre Theroset.

"It *must* be done; otherwise there will be a total breakdown of law and order," said Fernand Clerc.

The garrison troops, said Mayor Fouché, would insure that that would not happen: the previous night they had proved able to quell trouble.

"All the same it will be impossible to house and feed all the new arrivals," argued Clerc.

"Then action must be taken to insure that it is possible!" said Robert de Saint-Cyr.

The action was to be simple and far-reaching. The mayor could assume a wide range of powers should an emergency arise.

He would now invoke those powers.

He would insure that anyone with accommodation to spare would place it at the disposal of the refugees from the countryside. Food supplies would be brought by sea from Fort-de-France by all available boats, and by mule trains from villages in the southern half of the island.

Looting would be punishable by summary execution. No one was to be exempt.

"These are extreme times, gentlemen, and they call for extreme measures," concluded Mayor Fouché, using the clichés that men have always resorted to when assuming special powers over their fellows.

CHAPTER EIGHT

A Black Mass

WHILE THE "action committee" deliberated, the townspeople had been going to the polls to cast their ballots in the primary. The polling booths had opened in each of the town's quarters at eight o'clock; they would close at two P.M. and the result would be announced by Mayor Fouché from the Town Hall in the early evening.

Both the Radicals and the Progressives had made strenuous last minute appeals to the electorate, and the air was filled with conflicting slogans.

The finer issues of the conflict were forgotten in the last hours. Both sides were appealing to the man in the street. There was a reasonable grasp of the main issue: whether white or black power would triumph. The ballot had become a gladiatorial contest between an ineffective white man and two aggressive black men.

The three candidates had all voted early. Now they would spend the day in various ways. Clerc would go to the Cathedral of Saint Pierre to celebrate Mass, and afterward concern himself with business; the two Radicals would continue to whip up support right up until the moment the polling booths closed. They continued to exploit the behavior of Pelée; the volcano, as if signifying support for their cause, did not let them down.

In his study, Louis Mouttet had laid his plans to cope with Pelée. He would set up a Governor's Commission of Inquiry.

In silence Andréus Hurard listened as Mouttet outlined the details for Martinique's new Governor's Commission. Its president would be Lieutenant-Colonel Jules Gerbault, the Deputy Governor, and "chief of artillery on the island." The serving members would be Paul Alphonse Mirville, who was "head chemist of the colonial troops"; William Léonce, "assistant civil engineer"; Eugène Jean Doze and Gaston Jean-Marie Théodore Landes, both professors of natural science at the Lycée of St. Pierre. Landes, at forty-two, was a scientist whose reputation had spread far beyond Martinique. He would, reflected Hurard, give the Commission the status it badly needed.

Yet status would not be enough to save this Commission from the charges it would later face. Previous Commissions had taken months to sift evidence; serving members had been picked for known impartiality, and their appointments later ratified by the Government in Paris. But Mouttet, in a need "to effect domestic tranquility and overcome skepticism among foreigners," had decided to put aside tried and proven methods of appointing a Commission and its terms of reference.

The Mouttet Commission had the hallmark of the absurd stamped all over it. Apart from Landes, not one of its members was competent to offer any scientific estimate as to how Pelée would continue to behave. Instead of months, the Commission would have just two days to assess any threat to St. Pierre. On Wednesday, its findings were to be published in *Les Colonies*.

The Commission's purpose was clear to Hurard: it was to be a tranquilizer, a means of endorsing Mouttet's assessment of the situation.

The Governor had one other card to play in his political game. He would reprieve Auguste Ciparis, the Negro languishing in St. Pierre's military *bagne*. He would make the announcement at the civic dinner which the mayor of St. Pierre gave every year for the Governor on the eve of Ascension Day. The timing would give him the maximum political advantage, coming as it would on the heels of his Commission's report, and would insure the victory of a Progressive in the run-off election next week.

To Andréus Hurard this crude package deal was further proof of Mouttet's control of the situation.

The lack of control which Sequin, the master of the *Topaz*, had detected in the Governor the previous night had been placed under careful lock and key. To Hurard, the Governor appeared no more than "excessively tired. His face was worn and drawn from a night of little sleep, and he tugged frequently at a twitch under his left eye." The harshness and raucousness that Sequin had noticed were gone, replaced by a voice "that was no more than a whisper."

Mouttet explained to Hurard that he had spent most of the night alone in his study reviewing the situation. He had come to the same conclusion as before: Pelée offered no threat. This view had been strengthened by reports of the volcano's activities during the night: a burial party had been disrupted; a tidal wave had wrecked the power supply to St. Pierre; a number of mulattoes had been killed on the waterfront; pumice had rained on the town. But none of these incidents, alone or collectively, argued Mouttet, could justify the panic measures now being canvassed. Food might be scarce, along with accommodations. But that was no cause for spreading the alarm.

Five miles away, jogged by the gentle motion of his carriage, the Vicar-General of Martinique, Monseigneur Gabriel Parel, had decided upon his text for the sermon he would preach in the Cathedral of Saint Pierre later this morning. He would take as the basis of his sermon Psalm 46: "Therefore will we not fear, though the earth be removed, and though the mountains be carried into the midst of the sea. Though the waters thereof roar and be troubled, though the mountains shake with the swelling thereof. . . ."

The words stirred the Vicar-General. They fitted his own sentiments perfectly. No matter what nature unleashed through its agent Pelée, no harm would come to the flock of the acting head of the Roman Catholic Church on the island as long as they put their trust in the Lord.

A month had passed since the Bishop of the diocese,

Monseigneur de Cormont, had sailed from Fort-de-France for three months' meditation in France. In his absence the Vicar-General had assumed wardship for the island's spiritual needs, or at least as much as the Catholic Church could embrace.

This past month had been a trying time for the Vicar-General, and he hoped he had not been found wanting. The erupting volcano had cut deep into his routine. He had traveled to all parts of the island, offering reassurance, praying where disaster had struck, bolstering up his priests, promising deliverance to their congregations. He had taken innumerable soundings as to the danger that Pelée threatened. In the end he had come to his own conclusion: nobody really knew what was going to happen, but if people retained their faith in the Lord and in the power of the Church, all would be well.

At fifty-five, the Vicar-General had two outstanding attributes. He had an uncanny ability, once he had chosen a theme, of being able to deliver a lengthy sermon without any reference to notes; in a Faith that is not given to spontaneity, he had the easy flow of words more common to evangelists. His second asset was an ability to write; though the prose might be purple at times, he had a graphic eye for detail.

This Sunday morning only the opening chapters of the drama had been entered in the large notebook that the Vicar-General carried everywhere with him. Now, in his carriage on the way to St. Pierre, he read again the words he had witten—not, as he wrote later, "from personal vanity, but to insure that I have not neglected the Church in any way in this matter."

He need not have feared. Page after page of "impending catastrophe that has no parallel in our history" was filled with tales of people seeking sanctuary in the Church, praying, being prayed for, blessed, and prayed for again. The priests on the island, from the one in Le Prêcheur to those at the Cathedral of Saint Pierre, had behaved magnificently in sustaining the courage of the people.

These were, he had written, "trying times; one does not know when new disaster will visit us." But when it came, the Church would be ready, for "the Fathers are not among the last to realize the dangers of Pelée."

During the previous week the Vicar-General had traveled along this same road "in a downpour of ashes which spread a strong odor of sulphur. I visited places near the volcano, and on the journey I saw that the roads were filled with country people fleeing from the hills to the safety of St. Pierre, where with outstretched arms the people besought the priests for absolution."

Now the road was empty. As the coach began to climb the last slope before the road dipped down into St. Pierre, the Vicar-General came to the end of his chronicle, satisfied that the Church had not been neglected.

In the fields around the Guérin sugar refinery, the Vicar-General could see the workers picking their way in ragged lines through the cane. There could be no reason, except greed, for working on Sunday. He would draw attention to this in his sermon. The Sabbath was a day of rest.

In many ways the Vicar-General was an unusual man. Ten years had passed since he had come to Martinique, but he still managed to maintain an aloof distance from the intrigues, scandals, and political in-fighting that were the staple diet of many of his flock. Mouttet he knew slightly; Clerc was no more than another communicant to him; the Diplomatic Corps was only a block of seats in the Cathedral. He believed in a militant Church, but he did not believe that militancy should reach out into the secular intrigues of the island.

He had found friendship in the villages and hamlets, among "the true Martiniquians. They are kindly in spirit, the women more particularly, graceful and dignified in bearing, and both sexes sufficiently alive to the recognition of their worth. The men do not differ radically from other Negro types that are distributed through the Caribbean, except that they are softer in character and more gentle in ways. It is different with the women, who appear immediately as a race apart. Of unusual height, supple and straight as their royal palms, these proud products of Martinique at once arrest attention; and while one could readily challenge the contention that they are the fairest of the fair, it may be admitted that some of them are imperiously attractive, and that a voice more beautiful than theirs or one better qualified to charm cannot be found as a quality belonging to any other race."

Nowhere had this kind, gentle man found those voices more lovely and more enchanting than during the singing at the morning service he conducted in the Cathedral of Saint Pierre every alternate Sunday. And for all its raffish reputation, the town itself evoked a huge warmth in the Vicar-General. Now, from his vantage point, the buildings appeared to him to huddle together as if seeking reassurance under the pall of ash and cinder.

On impulse the Vicar-General closed his eyes and murmured a prayer for the safety of the town.

St. Pierre would need more than prayers to save it. It would need a crusading Churchman who would use his immense power to force decisions in the coming days. Instead, in the Vicar-General of Martinique, the town had a romantic ideologist who believed implicitly that prayer was the panacea for all troubles.

Three miles away, Dr. Guérin, the man the Vicar-General planned to censure in his sermon for breaking the Sabbath, was preparing to inspect his workers. Square in the saddle on his gray stallion trotting out toward the fields, he could hear ahead the steady beating of a drum and a low moaning chant. Then through a gap in the cane he saw the long double rank of workers going steadily forward, leveling the tall cane as they progressed, singing, accompanied by a drummer boy who kept abreast of them.

The men worked in pairs, each pair fifty feet from the next. Each man carried a murderous-looking weapon: a heavy bladed sword, the machete of Latin America, nearly four feet long and curved like a buccaneer's cutlass. The swordsmen wielded the machetes with both hands, swinging the heavy blades close to the ground, the tall stalks toppling before the steel in windows.

A few yards behind each pair of men came a woman, whose task it was to gather up the long stalks of cane and bind them in bundles. These women were the *ammareuses*, a Creole corruption of the French *amoureuses*, the lovers, a nickname for the favors they readily dispensed during the breaks for food.

Dr. Guérin watched as the bundled sheaves of cane were loaded on to clumsy wooden-wheeled carts drawn by oxen. Work was progressing well; his son Eugène told him the carts were on their third journey to the factory. He was about to turn toward the building himself when a sudden screaming came from the far end of the advancing ranks. The line wavered, then broke, as hundreds of workers rushed to the scene.

When Dr. Guérin reached the area, there was nothing anybody could do. The earth had suddenly split open and sucked twenty men and their women into the crack; then, freakishly, the gap had knitted together, crushing to death the cane cutters and their *ammareuses*.

From the surviving workers came a rumble of fear. Without waiting to free the corpses from the fissure, they scattered toward their homes, crying that Pelée had struck again.

In the Rue du Collage in St. Pierre, another crowd was running. It was in hot pursuit of a young Negro whom hunger had driven to snatch some food from a café. The owner had given chase. Others, possibly from boredom, or seeking relief from the misery of their own hunger, or just because it was a Negro being chased, had joined in. The Negro was halfway down the Rue du Collage when they caught him. He threw the stolen fish at them as he fell beneath their clubbing fists. The first sign of mob rule had come to St. Pierre.

In the Jardin des Plantes, the town's botanical gardens, Professor Gaston Landes was contemplating the latest damage. No longer would the gardens supply Paris, Berlin, or Kew Gardens with tropical plants. The palms, ravenalas, rubber trees, giant cacti, and red hibiscus lay dead beneath a shroud a foot deep in places. The ground was littered with birds who had been choked by the fumes and gases which Pelée had poured into the atmosphere.

The destruction was a personal misfortune for Professor Landes. Later in the year, during the Lycée's long summer vacation, he was due to sail to France with a selection of bulbs for the botanical gardens near Paris. But the bulbs, like all else in the Jardin des Plantes, were dead. The trip, financed by the Lycée, would have to be canceled. In a fit of disappointment, Professor Landes began savagely to uproot a bed of hibiscus.

The British Consul General, James Japp, was also behaving out of character this Sunday morning. He had informed his startled servant, Boverat, that for the first time since coming to the island, he would cook his own lunch.

In his room at the Lycée of St. Pierre, Professor Roger Bordier was writing a letter to his wife, convalescing in Paris after a long illness. He wrote slowly, frequently stopping to contemplate Mount Pelée, whose slopes rose from just beyond the school on the outskirts of the town. He had written to her every day since she had left the island seven months earlier. The letters were a *potpourri* of thesis, observation and trivia. He sent them, in bundles, on the weekly mail ship to France.

Tutor in history at the Lycée, he had come to an intriguing conclusion in the last few days. He had found a marked similarity between St. Pierre's predicament and the destruction of Pompeii by Vesuvius in the year 79. Until now he had told no one of his theory, fearful, possibly, of ridicule from his colleagues and pupils. But to his wife he confided the link he had found. Both towns were ports of the same size. Both stood in the shadow of their volcanoes. Both had been subjected to a lengthy fallout of ash, cinders, and gases. Both had experienced earth tremors. In the case of Pompeii, destruction had been sudden and swift.

"I do not say this could happen here; yet in these days anything can occur," wrote Professor Bordier with uncanny perception.

Only in one important fact would Pompeii differ from St. Pierre. In Pompeii nearly all the inhabitants had left the town before it was destroyed, a fact which is confirmed by the small number of bodies found in the ruins. In Pompeii no more than a thousand people died; most of the bodies that have been excavated suggest a total indifference to impending disaster.

It was the indifference of those victims that fascinated Professor Bordier; he strongly sensed its counterpart in St. Pierre now.

Professor Bordier was baffled by the people's attitude: "They have a blind faith in the protection of the town." But already that protective shield had been pierced time and again: "Everywhere, inside and out is covered with cinders. The floor, the furniture, it even penetrates into the drawers. When I go out into the streets, my hat is covered with ashes to a thickness of several millimeters; my alpaca vest is gray, my trousers and my shoes are the same color. In the gardens of the Lycée, the little birds hardly know on which branch to perch; the pigeons are cowering in their houses; in the yards choking hens and ducks remain in the coops. The appearance of the country is dismal. It is grayer than when it rains. In fact it is raining, only it rains ashes."

As he wrote, his eyes were distracted by movement around Pelée's summit. A finger of smoke had risen several hundred feet into the air above the crater. For a while it hung there, black and straight. Then, under pressure from inside the crater, the strand of smoke started to curl up on itself, flattening out at the top to form a small cloud. The cloud wavered and was then dispersed by the air currents raging around Pelée's head.

The whole phenomenon had lasted only a few moments, but to Professor Bordier it boded ill. He calculated that the column must have contained a high proportion of solid matter to remain suspended that long in the fierce air currents; yet the pressure building up inside the crater was enormous enough to have suddenly curled up the plume into a cloud and then scattered it. To his absent wife, after describing what he had seen, Professor Bordier posed one last thought: "Who knows what will be the next phenomenon, and how will we face it?"

He sealed the letter. It would be in the last batch of mail ever to leave St. Pierre, stowed in the hold of the Italian ship *Orsolina*, now anchored in the roadstead off the town.

A Black Mass

From the bridge of the French cable steamer *Pouyer-Quertier*, Captain Jules Thirion was observing the volcano and the landscape below it with care. He had sailed, under the instructions of Admiral Pierre Gourdon, Commandant of the French Naval Force of the Atlantic, from St. Lucia the previous evening. His instructions were to produce a chart of the sea bed in the area where the cable linking Martinique with Dominica was believed to have broken. It meant surveying an area from just south of the island to a point three miles north of St. Pierre. The work had started at dawn. Now the cable ship was approaching St. Pierre, a mile on its starboard side.

Down on the deck an officer of the watch supervised the essentially boring work of taking soundings and logging them. Later all the soundings would be transferred to a trace which would form the basis of the chart to be used in plotting the actual recovery of the severed cable from the ocean floor.

From the bridge of his ship, Captain Thirion watched the cloud puff over Pelée disintegrate. It had been another indication of unusual activity on the island since he had arrived. He logged it.

Seen from the sea, the island rose in a series of bold, rugged cliffs, its slopes covered with forest and fields of sugar cane. The lesser heights seemed to climb like huge camel humps until they merged with the mountain slopes of the hinterland.

The island reminded Captain Thirion of no other island he had seen: its gently swelling outline didn't call to mind the crags and cliffs of Capri, Ischia, or any of the other Mediterranean islands he had sailed past; nor did its heights recall the nearer mountains of Cuba, Jamaica, or Puerto Rico. The landscape was uniquely that of the Lower Antilles, creating its own atmosphere, one dictated by the towering Pelée.

Through his telescope he studied St. Pierre. The twin towers of the Cathedral rose majestically above the waterfront. In the roadstead nine ships were at anchor. In the morning light the town was quiet and peaceful.

For some time now as the *Pouyer-Quertier* had nosed

up the coast of Martinique, the sea had turned from dark blue to milky white. Now, off St. Pierre, it was a gray-white; floating in the ash was a wide variety of debris, dead animals, and human corpses.

An officer had asked if they were to try to pick up any bodies, and Captain Thirion had said they were not, because of the risk of contamination. He had instructed the lookout to give the helmsman ample warning so that he could steer clear of any carcass or corpse.

Captain Thirion was about to leave the bridge when he noticed a fresh puff of smoke being ejected from the crater. This time it shot several thousand feet into the air. Moments later another puff, larger than the first one, rocketed upward to join it. In moments a steady stream of black smut was funneling out of the volcano to form an ever growing cloud that hovered over Pelée.

Then slowly the cloud began to move, still growing, fed through the umbilical cord that linked it to the volcano.

All work on the *Pouyer-Quertier* stopped as the crew watched to see how this phenomenon would behave. They had not long to wait. Gathering speed every second, the cloud started to spread out over the sky like a giant black fan. In moments it was racing high over St. Pierre. It made no sound.

To Captain Thirion it was "as the twilight of an eclipse settling over the land and sea. The spectacle of the advancing ash cloud, like a huge octopus overshadowing all, caused us all to direct our gaze upward, full of fear. It seemed as if the end of the world had come. Except where illumined by the sun into a dazzling white border, its colors were a cold and forbidding gray-black. I roughly estimated the height of its course to be not less than thirty thousand feet above us, and it may have been more."

As the cloud passed over the cable ship, the funnel of smoke linking it to Pelée's throat died away, leaving a mass of several square miles racing across the sky.

But now, freed of its tie with the volcano, the mass appeared to be slowing down, as if it had run out of velocity. It also appeared to be losing height.

A Black Mass

As it did so, it suddenly disgorged a shower of stones and mud. The first fragments of pumice shattered on impact on the *Pouyer-Quertier's* decks, sending hot fragments flying in all directions. Several crew members were injured by the splinters. Others were singed and blistered by the hot mud.

"Acting on instinct, I ordered all speed to be made from the area," Captain Thirion later recorded in his log.

From the safety of the bridge house he watched the last stages of the phenomenon play itself out. A couple of miles away on his port side, the giant cloud was visibly slowing down. It lingered in the sky, a black, evil mass. Then it dropped, dipping and swooping in an arc, down to the surface of the sea.

"I trained my glass on the area, unable to believe what I was witnessing. The mass sent great clouds rising from the water long after it had disappeared, and I concluded it was steam caused by the heat from the cloud," Captain Thirion logged.

The great cloud had been no more than a curtain-raiser for what was to come.

CHAPTER NINE

The Prison Riot

THE VICAR-GENERAL saw that the cloud had brought panic to St. Pierre, and panic was something he knew how to cope with. He ordered his coach driver to make for the Place Bertin. The frightened rabble fell back as they recognized the carriage of the acting head of the island's Church.

"Make for the center of the square," shouted the Vicar-General, leaning from his carriage window.

His presence had a remarkable effect. Slowly the shouting and screaming died away until, in the end, the Place Bertin was totally still.

Only then did the Vicar-General step from his coach. Pulling his cape around him, he climbed up to stand beside the driver and looked out across the mass of expectant faces.

Of this moment he was later to write to his superior, the absent Bishop Alfred De Cormont: "They sought from me one thing only, reassurance. I could only think that if they sustained their faith, the Lord would protect them."

As the Lord's immediate representative, the Vicar-General now set about providing that reassurance and shoring up the faith that the cloud had threatened to undermine during its passage over the town.

"There is nothing to fear!" His voice, firm and strong, carried across the Place Bertin. "Look around you. There is little smoke, and what there is now blows away from the town."

The Prison Riot

A thousand faces turned to follow the Vicar-General's out-stretched finger. It was true. The wind had suddenly veered, the first time it had done so in twenty-three days. It was now blowing north, carrying a thin trail of smoke in the direction of Dominica.

"Let us pray," ordered the Vicar-General. "Let us pray that God will deliver us from further evil."

For a moment no one moved in the square. Then a man standing near the carriage dropped to his knees. The move was a signal for people all over the Place Bertin to adopt the accepted posture for prayer.

"Gloria in excelsis Deo. Et in terra pax hominibus bonne voluntatis. . . ."

The patter of the Gloria in Excelsis carried around the square: "We praise thee, we bless thee, we adore thee, we glorify thee. . . ."

Slowly the crowd picked up the chant: *"Gratias agimus tibi propter magnam gloriam tuam; Dominus Deus, Rex caelestis, Deus Pater omnipotens. . . ."*

To Clara Prentiss, making her way to the Cathedral for morning Mass, the chanting "was the most glorious sound I have heard." Drawn by it, she made her way to the Place Bertin. There, on the edge of the square, she too knelt in the dust, heedless for once of the ash, to join in worship.

"Domini Fili unigenite, Jesu Christe. Domine Deus, Agnus Dei, Filius Patris. Qui tollis peccata munde, miserere nobis. . . ."

Andréus Hurard heard the chanting when he was still several streets away. He had ridden hard from Fort-de-France, eager to get back to his office. There, every Sunday morning, he composed the main editorial for Monday's issue of *Les Colonies*: it was always a reflective one, looking back at the weekend which had passed, pointing up what the week might contain. Usually, he devoted the space to a political issue. But now, he had decided, the space would be best used to bring into perspective the threat of Pelée.

The sight in the Place Bertin emphasized to him

"how the whole issue had been magnified. There could be no justification for such prayers."

Dismounting from his horse, he started to lead the animal around the fringe of the crowd now coming to the end of the Gloria in Excelsis: *"Tu solus Dominus, tu solus Altissimus: Jesu Christe, cum Sancto Spiritu: in gloria Dei Patris."*

The "Amen" was lost in a sudden cry: "The volcano! Look at the volcano!"

Huge cauliflower-shaped yellow clouds had started to roll out of the crater. The crowd rose from its knees, still silent, awe-struck by the spectacle. In his notebook, Andréus Hurard recorded for posterity the sight:

"We turned our eyes in the direction of Pelée, and the sight that met them was truly terrifying. The crater, peaceful until a moment ago, was hurling out wild sheets of yellow cloud. They came rolling and puffing with great fury, and in an instant the sky to the north was filled with smoke that shifted and rose, twirling itself into lofty columns and pyramids or mushroom caps—rolling black and yellow with the angry ashes that were being carried out by them. Pelée's top was being lashed in fury by the smoke and was soon buried in the dark shadows which this new example of life in the crater had called forth. The scene was an extraordinary one, made doubly impressive by the rapidity with which it was brought about."

The eeriness of the phenomenon was enhanced by the total absence of noise that accompanied it. In his notebook, the editor added: "Soon it became obvious to even the most timid that the clouds presented no threat." Later, in his office, he expanded this theme for his editorial.

In his opening sentence, Hurard summed up the behavior of Pelée over the weekend. "The people of St. Pierre were treated again to a grandiose spectacle in the majesty of the smoking volcano." He started to analyze the exact content of the ash falls, finding comfort in the fact that they contained "neither lime nor sea salt, nor any chemical substance that could be injurious to vegetation." Turning to the damage the ash had done, he placed the blame on the weight of the falls, not on their content, adding: "While the branches of many bread trees have been broken, it must be remembered that the wood of a bread tree is of course very fragile."

Finally he turned to one last question: "Shall we have further earthquakes? It is not probable." For "men competent to judge" had informed the newspaper of this fact.

These men, if they existed at all outside the editor's mind, have never been traced.

Out in the roadstead, Captain Marino Leboffe of the *Orsolina* had come to a decision. He would, if necessary, sail with his holds empty rather then risk being engulfed.

There was silence in his cabin as he announced the decision.

The *Orsolina* was one of the four Italian ships riding at anchor off St. Pierre. There was the *Teresa Lo Vico*, a wooden bark of 585 tons, waiting to take on board a cargo of sugar from the Guérin refinery. Another bark, the 580-ton *Sacro Cuore*, had just arrived from Marseilles with a mixed cargo. The *Nord America*, a 583-ton bark, was waiting to load a mixed cargo from the Clerc warehouses.

At midmorning the captains of the barks had assembled aboard the *Orsolina* to discuss the eruptions from Pelée.

They were reluctant to endorse the decision Captain Leboffe had come to. They argued that to sail without instructions from their owners would be in breach of their agreements.

"Besides," said one of the bark captains, "Pelée is not Vesuvius."

"Indeed it is not. I know what Vesuvius is. And I feel that Pelée is much that Vesuvius is not," replied Captain Leboffe. "If the volcano worsens, I shall turn my stern on Pelée whatever the consequences!"

In St. Pierre, in spite of the efforts of the Vicar-General, fear still ran high. In the mulatto quarter, Father Alte Roche failed to prevent a group of families from leaving their homes and boarding their fishing smacks to sail down the coast to Fort-de-France. Other groups were setting off to make the journey on foot. Near the British Residency—where James Japp prepared lunch—

they encountered the survivors from Ajoupa-Bouillon. Confusion and panic spread, carrying with it wild rumors to all parts of the town.

With the rumors came an increase in the pillaging. By the end of the day, thirty-seven separate incidents of looting from food and vegetable shops would have been reported. The police, who with the local garrison would have found they were almost powerless to control the outbreak, would receive instructions from Mayor Fouché, under "powers invested in me in a time of emergency," on how to cope with the situation. Looters were to be arrested, tried on the spot, and if found guilty, confined in the military prison "until further notice." Rough justice had come to St. Pierre; it would do little to reduce the looting of food and fruit; it would do nothing to reduce the fear that was now eroding the orderly commerce of the town.

The "action committee" had organized convoys to travel to all parts of the island to bring supplies in to St. Pierre. Even before they had left the town, the convoys found themselves brought to a standstill by parties of refugees coming into and leaving St. Pierre.

To Thomas Prentiss, the American Consul, watching the scene from his balcony, "the over-all impression is one of total and utter chaos. Everyone in sight appears aimless. Law and order as we know it is on the decline. If help from outside does not come soon, I fear for the consequences."

But the world outside Martinique had not yet heard one word of what was happening on the island.

The tolling of the bells in the twin towers of the Cathedral of Saint Pierre woke Auguste Ciparis out of a fitful sleep. When the bells stopped, it would be midday. High Mass would begin, as it had begun every other Sunday in the Cathedral. Ciparis listened intently to the ringing bells, aware that unless the miracle he prayed for happened, it would be the last Sunday he would hear them. The bells would not ring again until after breakfast on Thursday, Ascension Day; by then he expected to be dead, hanged by the neck on the gallows that would rise from the wreckage of the platform in the corner of the prison yard.

The Prison Riot

The *bagne* of St. Pierre was one of the few that still retained the gibbet for executions. In metropolitan France the instrument invented by Dr. Guillotin had come into wide use, replacing the firing squad which in the martial spirit of the First Empire had been the accepted method of execution. The firing squad had never found favor on Martinique, simply because there was a shortage of skilled marksmen among the guards. The guillotine, in spite of theoretically being a swift and precise way of killing, had been rejected because the prison administration believed that the blade and framework containing it would warp in the tropical climate, producing what one report on the matter called "terrible consequences. Even if the guillotine was in perfect condition, if its uprights were not checked with a plumb line to see they were absolutely vertical, the knife would not run smoothly and would stick on the way down, cutting into the neck of the condemned man without completing its task. This would be distressing"—not least, presumably, to the condemned man.

While humanitarian principles had gained ground from year to year, especially after the Empire, the discipline in a *bagne* such as St. Pierre's remained harsh. Sunday in particular was hard. Officially ordained a rest day, in reality, it meant the prisoners were confined to their cells except at mealtimes.

Since the doors had clanged shut after breakfast, the *bagne* had been a silent place. It had been broken by a chorus of fear from the prisoners as the ash cloud had cast a dark shadow over the town; another tumult had followed the appearance of the soaring cauliflower clouds pouring out of Pelée. The outbursts had been quelled by the guards on duty banging on cell doors with their musket butts. For a while silence had returned.

But now, as the bells of the Cathedral stopped, Ciparis heard a fresh sound emerging from the prison. At first it was no more than a whisper, falling away, then returning a little stronger than before. One moment it seemed to come from somewhere in the cellblock above, the next it was coming from a block on the far side of the prison; then the gap was bridged, and the sound, low and guttural, came from all parts of the *bagne*.

The prisoners were demanding to be transferred to the *bagne* at Fort-de-France until Pelée had ceased its threatening behavior.

For the first time in the history of the French colonial penal system on Martinique, a riot was underway. Fanned by the harsh conditions and the fear and the suspicion, it would grow rapidly unless quelled at once.

Nothing disturbed the solemnity of High Mass in the Cathedral of Saint Pierre. With one exception, the Cathedral was completely filled for the great Christian sacrifice of the Catholic Church. The exception was the empty pews normally occupied by the Guérin family. Their absence had caused excitement among the congregation. The Guérins had never been known to miss Sunday Mass. Shortly before the service commenced, news of the tragedy at Ajoupa-Bouillon rustled through the congregation.

But in other parts of the Cathedral, there were others with happier thoughts on their minds. René Cottrell, aged twenty-three, was looking forward to the end of Mass when he would escort the pretty, young girl beside him back to his uncle's house for lunch and a leisurely afternoon of talking and reading.

René Cottrell, in St. Pierre only a week, had already fallen in love with Colette de Jaunville.

The Cottrells were one of the families in the Ten, with estates around St. Pierre and Fort-de-France. The de Jaunvilles also had extensive plantations, and could trace their aristocracy back to Josephine Beauharnais, who married Napoleon.

On an island where it was difficult to arrange suitable marriages, this one would be smiled upon. René Cottrell was looked upon as one of Martiniques's most eligible young men. Colette de Jaunville—like her distant ancestor, the Empress Josephine—was an affectionate, talkative girl, with thick, curling chestnut hair and dark eyes, fringed with long upswept lashes. She was fifteen, a little past the age when many girls on the island would expect to be wed. But now she could look forward to life as a rich planter's wife, exercising her charm and hospitality on the petty society of Martinique.

Now they knelt together for the words of the Consecration.

Professor Bordier, seated at the back of the Cathedral where by tradition the staff of the Lycée sat, was also following the words of the Consecration with particular interest. Before the Mass he had confided in Professor Landes his fear that if conditions worsened, it might be impossible to conduct further services; people, he had said, would not be able to pray in the sulphur laden air. Professor Landes had dismissed the notion as "beyond my comprehension." He had then told Professor Bordier that he had been asked to join a Governor's Commission of Inquiry into Pelée, adding, so Bordier wrote later, the significant words: "It is clear there are too many wild rumors abounding." Certainly any suggestion that the Cathedral services should be suspended were, on this Sunday morning, not in keeping with the situation: the air in the Cathedral was relatively free of sulphur or ash particles compared with most other public buildings in the town, though Clara Prentiss noticed "a thin film of gray was everywhere giving the Cathedral a funereal appearance."

Now, as the Mass commemorated the dead, she could not help feeling that soon there would be more, new dead to remember.

Dr. Guérin, with the help of his sons Eugène and Joseph and men from the factory, dug out the bodies of the twenty workers where they had been crushed. They laid them out on a wagon, and then drove to the burial patch behind the factory. The estate workers stood in weeping groups. As the last shovel of earth was replaced, Dr. Guérin offered prayers for the dead.

In his church in the mulatto quarter, Father Alte Roche was coming to the end of morning Mass. Unlike the Cathedral, where the Mass had been sung, the service the Jesuit conducted was a simple one, with a mulatto boy acting as server. And this church lacked the rich tapestries, gold, and candelabras of the Cathedral. It could hold about three hundred people; every seat had been filled before the Mass started.

Father Roche was giving the congregation the blessing when a rattle of musket fire cut across his words.

Eyes strayed uneasily to where the sound had come from. For a brief moment Father Roche faltered, then picked up the words again; ". . . *et habitavit in nobis; et vidimus gloriam ejus, gloriam quasi Unigeniti a Patre, plenum gratiae et veritatis."*

From the kneeling congregation came the murmured *"Deo gratias."*

The Mass was over. As the congregation stirred to its feet, another crash of musket fire echoed in the air.

This time they were in no doubt as to where it came from—the direction of the *bagne*.

As the second volley of musket fire echoed in the prison yard, Auguste Ciparis felt a chill of fear. The guards, a dozen of them, deployed in a ragged line across the yard, were firing perilously close to the condemned cell.

From the block above the cell came the sound of wood splintering against metal. The prisoners had started to break up the funiture in their cells. Twice prisoners in other cell blocks had been quieted by the muskets. Now, a third round was poured through the barred windows above. For the first time in 143 years, the sound of concentrated gunfire had come to St. Pierre.

Out of the cellblock came the sound of men in pain. The splintering ceased, and from the yard came the voice of the prison governor.

Through his window slit, Auguste Ciparis could just see him: a middle-aged fat man.

The governor was brief and to the point. Unless the mutiny ceased at once, there would be massive reprisals. He was forthwith rescinding all privileges from the *éprouves*—the trusties; all prisoners would be on *grand fatigue,* and the ring leaders, once they were caught, would each receive fifty strokes of the *bastonnade* followed by solitary confinement.

This was one of the severest sentences listed in the penal code. The *bastonnade* was a six-tailed scourge, younger brother of the English cat-o'-nine tails. Each "tail" was three-tenths of an inch thick and eighteen inches long. It was attached to a thick wooden handle. *Bastonnade* punishment was carried out on a plank named the "justice bench." The blows were delivered from left to right across the shoulders. Solitary confinement was almost as severe a business. A prisoner would be tied to a wooden plank, with just enough play for him to be able to roll off the thwart to relieve himself.

In the *bagne* nobody moved. Then, from a distant block, Ciparis heard the sound of a cell door opening. Shortly afterwards a shuffling line of prisoners, escorted by armed guards, entered the yard. Soon another line appeared. Then, from above, he heard the bolts being drawn.

The riot was over.

CHAPTER TEN

Released to the Living

THREE MILES AWAY, on the slopes of Pelée, bad luck had struck the food foraging party from Morne Rouge. For three weeks the village had suffered from regular bombardments of ash which had contaminated the food and water supply. This Sunday morning after mass, the village priest had asked for volunteers to travel to the far side of the island to obtain victuals. Now as their mud-spattered wagonette was being pulled over a stretch of broken ground, it had broken an axle.

They continued without it, packing their way around massive boulders of basalt and diorite and climbing over ridges of ancient lava flows which linked together the volcanic masses scarring Pelée's slopes.

Above them the volcano was rolling out a volume of cloud and ash that fairly bewildered their senses. Two miles and more, the column of white curling vapors was climbing—lifting, rolling and unrolling, until it lost itself in the general obscurity as the arching vapors thinned out and melted into the sky. No sound came from the mountain. Below, they could just make out the noise of surf rushing against rocks, and nearer, the rustle of leaves as they tried to shed their crusts of ash.

By early afternoon the party had emerged upon the lower slope of the volcano and began traversing a long ridge about two thousand feet above sea level. Its gray and desolate surface, which until a month ago had been covered with grass and forest, was scarred with scoriae, small boulders, and sharply pointed pieces of rock, all of which had been ejected by the crater. For a moment they stopped to take a bearing. The ridge, wide at this level, contracted into a fairly narrow point as it traveled up to the summit. Below them the landscape receded, and beyond, the blue ocean dashed its white foam against the vertical cliffs of the coast. In the middle distance the Blanche River began its course; a muddy flowing sweep of chocolate that brushed beside the Guérin factory on its way to the sea. Morne Rouge, and in the distance, St. Pierre, appeared to be a collection of toy buildings from this height. Ahead and just above the ridge was a deep ravine, its walls like burnt scars, which had been cut by a mud torrent of some bygone era. Beyond the ravine lay the route to La Lorraine. Unless the party retraced their steps and tried to cross Pelée's flank from another direction, they had no alternative but to climb the ridge and hope that higher up the mountain the ravine narrowed enough for them to leap the gap.

The first stage of the ascent was an easy one; the ridge offered plenty of footholds for men brought up to scale peaks. As they climbed, the weather, which had been clear all day, started to deteriorate. Pelée's neck was ringed with thick mist that blotted out the dividing line between land and sky. Shortly afterward it started to drizzle; the eeriness was heighened by shifts in the mist through which they caught ghostly glimpses of *mornes* and *pitons*, tall, rocky spikes of rock growing out of the spine of the volcano.

The higher they climbed the greater became the signs of Pelée's devastation. Not a trace of vegetable growth remained. Then from above came a dull rumbling. Petrified, they stood still: now the noise appeared to come from somewhere deep below their feet; a moment later it seemed to emanate from the sea of clouds above them.

Suddenly a crash of thunder that seemed to rend the very heart of the mountain silenced all other sounds. An instant later there came a second clap of thunder, this time accompanied by lightning that cut frenzied slashes through the clouds. Then a third and a fourth crash echoed like artillery fire.

The foraging party were caught up in the crossfire of one of the massive thunderstorms which from time to time set Pelée's summit quivering. In a moment the rain was descending in merciless torrents, and the lightning cut blinding flashes, illuminating patches of the mountain for a brief moment.

Stunned by the force of the storm and soaked to the skins, the party started to slither and slide back the way they had come. It was a difficult and dangerous journey. The storm had washed away footholes, replacing them with new hollows and fissures. From these holes raced streams of mud and water, turning the hard slope of an hour before into a perilous morass.

They were halfway down the ridge when the lava flow caught up with them. Screaming with terror they tumbled down the ridge, the red-hot mass licking at their heels. Suddenly one of the men skidded, sending two of his companions sliding in the mud. Helpless, the three slipped over the side of the ravine, from whose depths curled puffs of vapor.

The rest of the party reached the end of the ridge and the comparative safety of an outcrop of rock. Below them the lava streamed past, smoky and glowing, slithering over the ground, its far bank from time to time dropping into the gorge.

Nearly maddened with fear, the survivors picked their way back down the mountain tracks, slipping and sliding most of the way down.

They did not feel safe until they had reached their damaged wagonette; by then the storm had started to lift. When it did so, it had carried with it the cauliflower clouds, leaving Pelée's head bare to the sky.

Below, the land was bathed in bright sunshine.

In his office, Louis Mouttet and the five members of the Governor's Commission of Inquiry were watching the storm blowing itself out around their subject's summit. This afternoon the Commission would travel to St. Pierre to begin their investigation.

"Will we have to travel to the mountain?" asked Paul Mirville, chemist to the island's garrison.

There would be little point, answered Professor Gaston Landes. He, for one, was well acquainted with Pelée's geography. He had a number of published topographic reports on the volcano in his room at the Lycée of St. Pierre which he would be pleased to lend to his fellow members if they needed to acquaint themselves further with Pelée.

The Governor had one observation to make: he wanted to emphasize that the Commission's role was principally "to establish the safety of St. Pierre," and not to discuss the mountain. Professor Landes had informed him that any serious lava flow would travel down the barren northern slope of the volcano.

"Your concern is to see how long St. Pierre can scientifically stand the strain of the ash and sulphur," explained the Governor. Their role was to see whether the terrain around St. Pierre would adequately shield the city from the danger of a lava flow.

It was, as others were to point out later, a curious assignment. For the Commission to virtually ignore the forces struggling inside Pelée and yet hope to arrive at a "scientific assessment" of the safety of the town it so clearly threatened was, to say the least, astounding. For Professor Landes, a man of known scientific integrity, to allow himself to be associated with such a brief was even more bewildering. Alone among the Commissioners, wrote Professor Heilprin later, Landes must have "fully realized the geological relations then existing."

If, in fact, he did, he kept them to himself.

Lunchtime had passed in St. Pierre. James Japp, the British Consul, enjoyed the meal he had cooked, stewed beef and potatoes. His servant, Boverat, had been surprised at the measure of his master's culinary abilities. Now, in their separate quarters in the Residency, they slept. Sleep had also come easily to Thomas Prentiss; years in the tropics had conditioned him to accept the siesta period as essential to a well-ordered way of life.

His wife Clara, unable to sleep, occupied herself with the island's other Sunday afternoon pastime, letter writing. Seated at a desk in the drawing room, she wrote to her sister Amy in Melrose, Massachusetts: "I write under the gloomiest impressions, although I hope I exaggerate the situation. Thomas laughs, but I sense that he is full of anxiety. He has stopped telling me to leave, knowing that I cannot go alone. The heat today is suffocating. When we came out of Mass, I saw that many of the people were obliged to wear wet handkerchiefs to ward off the sulphur fumes. Even though there have been no fresh ash falls for some hours, the air is oppressive. Your nose burns. I ask myself if we are all going to die asphyxiated. I wonder what tomorrow will bring.

"There are rumors everywhere of impending disaster. People talk of a flow of lava, a rain of stones, or even a cataclysm from the sea. Who can tell? Though for myself I do not believe all I hear. The volcano continues to surpass itself. After producing a great black cloud, evil yet impressive, it disgorged huge shapes of cloud which were lost in a storm. But in the last hour it is calm. My thoughts now turn to other things. Food is scarce. This lunchtime the cook could only offer chicken, beans, and potatoes. It was poor fare for a Sunday, but probably better than most could expect. There is talk of food coming into the town from other parts of the island, and this morning, on the way to Mass, I saw the carts leave. It is the first practical move by the Action Committee.

"The atmosphere in the town is strained. There are outbreaks of stealing and fighting. Troops are on hand to keep order, and Thomas says they are succeeding. Yet most people, in spite of it all, are content to stay in the town. This morning there was a small exodus from here, but now it has stopped. People sleep where they can, in the streets, even against the walls of our home. They wait, as we all do, for the arrival of the Governor's Commission. It is a curious thing, but I cannot share their relief at the coming of the Commission. How will the Commission end the dust that enters everywhere, burning our faces and eyes?"

She had raised a question that others would not ask until it was too late.

In other parts of the town, other letters were being written. Emile le Cure, the General Manager of the English Colonial Bank, was writing to a cousin in Marseilles: "Business in the past week is poor, yet my calmness astonishes me. There have already been a number of deaths caused by Pelée, and if I am to die at its hand, I am awaiting the event with tranquility. If death awaits me in this place, I will be in numerous company to leave the world. Few people talk publicly of the end. Certainly I do not encourage it among my workers. It would be easy to spread panic. And yet to myself I wonder how long it will be before the end comes. It will be what God wills. Tell Robert we are still alive. This will, perhaps, be no longer true when this letter reaches you."

The fatalism which had started to infiltrate St. Pierre is revealed in another letter written that Sunday afternoon. The writer was Véronique Clerc. To a sister in Lyon she ended a letter with these words: "If there is time, I will write and give you my last thought if I must die. But it is possible that death will be swift and unexpected. We are all prepared, and place ourselves in the safety of the Lord."

By early evening, by land and sea, the first supplies of food were on their way to St. Pierre. On board the *Topaz*, its master Jules Sequin was in command of a cargo of chickens, ducks, a litter of pigs, and sacks of fruit and vegetables. Around him a flotilla of small boats, the coastal steamers of the Girard Company, fishing boats, and sailboats, were all making their way out of the harbor of Fort-de-France. From the bridge, Captain Sequin watched the feverish activity of loading going on around the cable ship *Pouyer-Quertier*. Farther out, a couple of French naval ships rode at anchor, their sheets drooping from the yardarms like linen in a Neapolitan alley. Schools of gars and dogfish swirled about their hulk.

Sequin had been astonished at the speed with which the operation had been mounted.

He had barely finished lunch on board when a harbor official had arrived on deck with the news that the *To-*

paz—"in the name of the Governor"—was to sail with supplies for the relief of St. Pierre.

"Why not bring the people here?" Sequin had asked.

It was, said the official, "impossible." He added that the Governor was rumored to have given orders that "at all costs the population of St. Pierre was to stay in the town." To Jules Sequin, this was one more strand of evidence that Mouttet was going mad.

By midafternoon supplies of food, fruit and vegetables, were stacking up all along the waterfront at Fort-de-France. They were loaded directly into the smaller boats pulled into the jetties and ferried out to the *Topaz* and *Pouyer-Quertier*. There had been several lively discussions about who was going to pay for the goods; in the end the matter had been resolved by allowing a merchant to travel with each shipment and be responsible for selling it in St. Pierre.

Now, laden to the gunwales, the *Topaz* made steam for St. Pierre.

From villages in the southern half of the island, the food carts, loaded with produce, were also trekking back to St. Pierre. Between them, ships and carts had virtually cleaned out every storehouse. By nightfall there would be enough food in St. Pierre to last a week.

"After that we will have to make other plans. If need be, we will bring supplies from other islands in the Caribbean," Robert de Saint-Cyr told the "Action Committee," as it reviewed the situation later that evening in the members' lounge of the St. Pierre Sporting Club.

In the *savane* of Fort-de-France—the great green public square with its tamarinds and *sabliers*—Louis Mouttet stood before the statue which had been erected in marble memory of the Empress Josephine.

Over the years sea winds had worn it; tropical rains had streaked it; dirt had darkened the exquisite hollow of the throat. Yet the figure had retained a curious human charm. Her sculptors had robed her in the fashion of the First Empire, with gracious arms and shoulders bared, one hand resting upon a medallion bearing the profile of Napoleon. Her eyes looked back to the place of her birth, back to the dreamy village of Trois-Islets.

The statue, set in a circle of seven palm trees, had lately come to fascinate the Governor. During periods of acute stress, he would adjourn to the peace he found within the circle of palms. Looking up to the broad chin of the Empress, the Governor would stand there, an earthbound Napoleon whose mind preferred fantasy to fact.

To the Vicar-General returning from St. Pierre, there was something pathetically sad and lonely in the sight of the island's chief administrator deep in communion with the statue.

In his diary that evening he wrote: "I debated whether to divert my carriage and wait for the Governor as he left the *savane* to give him the latest news from St. Pierre. But I decided to leave him to himself."

The Vicar-General could be forgiven. The fifteen-mile ride over a bad road had been tiring. Besides, there was little chance that Louis Mouttet would have welcomed yet another eye-witness, however exalted, who would bring further bad news.

In St. Pierre, evening was approaching. High over the town rose Pelée, heavily coiffeured in cloud shot through with lilac and purple: a magnificent Madras headband formed by the sinking sun.

On the brow of the hill, where earlier the Vicar-General had paused, Professor Landes and the other members of the Governor's Commission of Inquiry had halted to inspect the countryside.

Then, as the evening light made its slopes appear to undulate against the northern sky, the party began its descent into St. Pierre, turning frequently to look from their carriage at *La Pelée*.

The *mornes* were now in shadow. Their curious conical forms spoke of volcanic eruptions millions of years before, and the *pitons*, volcanic upheavals of splintered strata, looked like a row of church steeples. But the bulk of Pelée, a huge mass fifty miles square at its base, rose clear in the evening light: a succession of strange jagged ridges, terraces crumbling into other terraces which broke into ravines, here and there bridged by enormous buttresses of basalt, an extravaganza of lava shapes pitching and cascading into sea and plain.

Fifty-one years ago this Sunday, Pelée had erupted. Le Prêcheur then, as now, had been the first village to suffer. The villagers had been choked by the oppressive stench of sulphur. Then, on August 4th, 1851, "much trepidation was caused by a long and appalling noise from the mountain"—a noise compared by those who heard it as similar to the "hollow roaring made by a packet blowing off steam, but infinitely louder."

Professor Landes, riding beside the coach driver, knew by heart the sequence of events that had followed. The hollow roaring had continued through the night, deepening at times into a rumble like thunder. Next day St. Pierre had filled with cries of "The mountain is boiling!" The mountain was not actually on the boil, but panic had seized the population. On the morning of August 6th, 1851, St. Pierre presented a strange aspect. All the roofs, trees, balconies, awnings, and pavements were covered with a white layer of ash. The same showers blanched the roofs of Morne Rouge, all the villages about, and the neighboring country. The mountain was sending up columns of smoke or vapor, and it was noticed that the Blanche River, usually of a milky color, ran into the sea like an outpouring of ink, staining it black for a mile.

A committee appointed by the major of St. Pierre had been asked to prepare a report on the mountain's behavior. They found that a number of rents had either been newly formed, or suddenly become active, in a flank of the mountain. All were well clear of St. Pierre. But to reassure the population still further, members of the committee had lowered themselves down into the vents—"and it is noteworthy that their researches were conducted in spite of the momentary panic created by another outburst."

As his carriage entered St. Pierre, Professor Landes was detemined that members of the 1902 Commission would not indulge in such antics.

In the end, in spite of their "most strenuous efforts,"

Released to the Living

the investigation of 1851 had found that there had been no cause to fear. There had been no structural changes in the mountain; the roaring had been produced only by the violent outrush of vapor and ashes from the rents.

Professor Landes was prepared to be surprised, but he doubted if the Governor's Commission would find a dissimilar situation to report.

"Take away the ash and the sulphur in the air, and the actual damage caused by Pelée would so far appear to be small," he said to Paul Mirville as their carriage made its way through streets covered with dust.

For a scientist with a respected reputation in America and France, trained to evaluate facts and to pass judgment only after the most rigorous of inquiries, to have made such a preliminary assessment is surprising. Undoubtedly it contributed to what Professor Heilprin was to label as an over-all "lapse in judgment" by Gaston Landes.

During the eruption of 1851, Pelée had expelled no more than white vapor along with the showers of ash. But already it had now ejected far more sinister looking mucus from its throat.

In 1851, in hope of allaying the general alarm, a Creole priest had climbed the summit of the volcano and planted a cross on its crest. The flamboyant gesture had worked. But it would need a far greater act of faith now to remove the fear many people felt.

Certainly one group of men was not prepared any longer to put their trust in the town. They were the fifteen hunters who, early on Saturday morning, had fled from the lava which had trapped the wild hog they had been hunting. After the first flight of panic had subsided, they had picked up the parcels of boar flesh which they had left on the far side of Pelée and made their way in a wide sweep down into St. Pierre. They had sold the meat long before they had reached the mulatto market and had found that the money they made could buy little other food.

They had spent a hungry night of fear and no sleep. The behavior of the volcano this Sunday morning had only increased their fears. Ravenous and tired, confused by the sporadic fights over food that had broken out like ulcers all over the town, the hunters had decided to return to the slopes of Pelée. At least in the *grands-bois,* the primitive forest that from St. Pierre looked like a band of moss belting the volcano, they would be on familiar ground.

In single file they padded out of St. Pierre, passing on the way the carriage bringing in the Governor's Commission.

Outside St. Pierre's Town Hall, a large crowd had gathered to await the results of the ballot. Shortly after eight o'clock, Mayor Fouché, the three candidates, Senator Knight, and assorted officials of the Progressive and Radical Parties appeared in the entrance of the Town Hall. The crowd fell silent as the mayor stepped forward and announced the result: Fernand Clerc had polled 2,367 votes; Alfred Percin had polled 1,910 votes; Pierre Lagrosillère had polled 1,260 votes.

The difference between Clerc and Percin was not enough; they would go forward to a second ballot in a week's time. But, as Amédee Knight forecast, the outcome had already virtually been decided. Between them the two Radicals had collected 3,170 votes; now that bloc would give them a comfortable majority of over 800 at the final election. That majority could be further increased if any of the 747 abstainers could be persuaded to the polls.

All day Suzette Lavenière had listened to the activity going on outside her room. By midday, when food was brought, she had known the cause; the hospital staff were coping with the injured from Ajoupa-Bouillon. Late in the afternoon a tired-looking doctor had unlocked the door of her room. With him was Father Alte Roche, a friend of the Lavenière family for several years.

The Jesuit had unstrapped her, explaining that the hospital authorities now realized they had made an error in diagnosis, and that Suzette was no more insane than he, or the doctor.

Even for the persuasive Father Roche, the freeing of Suzette Lavenière was one of the most remarkable episodes of this Sunday. After morning Mass, a worker from the Lavenière estate had approached the Jesuit, explaining that he had come to St. Pierre fearing for Mademoiselle Lavenière's safety. After hearing what had happened, Father Roche had gone to the hospital, seen the senior doctor, and persuaded him to release Suzette from the asylum wing.

Whether the doctor would have agreed in normal circumstances will never be known, but these were not normal circumstances. Every bed was taken, and the emergency cases from Ajoupa-Bouillon were lying on *palaisses* in the corridors. The doctor, like all the staff at the hospital, had been overworked for many hours. Faced with a priest whom he knew to have influence with the hospital administration and who now demanded that a patient who appeared to be physically sound—shock was an unknown medical diagnosis in the island's medical service—be handed over to his care, the doctor took the easy way out: he agreed to release Suzette Lavenière, female patient 273. The bed she had occupied could be used for one of the acute post-surgery cases from Ajoupa-Bouillon.

Then Suzette herself had posed a new problem. She had been brought to the hospital as a stretcher case; she had no clothes with her, and could hardly return to the estate in a nightdress. The difficult had been resolved by the doctor's lending her the official nursing habit of the hospital—a long, clinging white smock.

Dressed as a nurse, she left the hospital with Father Roche as dusk approached.

Louis Mouttet spent the last hours of this Sunday drafting a telegram to the Minister of Colonies in Paris:

AN ERUPTION OF MOUNT PELEE HAS TAKEN PLACE. LARGE QUANTITIES OF ASH WERE SHOWERED ON THE NEIGHBORING COUNTRYSIDE. THE INHABITANTS HAVE HAD TO ABANDON THEIR DWELLINGS PRECIPITATELY AND SEEK REFUGE IN ST. PIERRE. THE ERUPTION APPEARS TO BE ON THE WANE.

This telegram, sent twenty-six days after the volcano's awakening, was the first official notice the world was to receive that Pelée was in eruption.

The telegram itself made no sense. It was prepared at an hour when Colonial Governors were not normally on duty unless a serious crisis threatened; and clearly, since "the eruption appears to be on the wane," Louis Mouttet did not appear to believe there was any threat.

There was no mention of the formation of the Commission of Inquiry, an oversight which compounded Mouttet's questionable action in setting up the Commission in the first place without the consultation with the Ministry required by his terms of office. Furthermore there was no reference to the formation of the citizens' "action committee" in St. Pierre, a serious extension of local government which was itself some measure of the gravity of the situation.

Most importantly, Mouttet's report in no way conveyed the widespread devastation and loss of life resulting from Pelée's activity. By this Sunday night there had been an almost complete destruction of the villages of Le Prêcheur and Ajoupa-Bouillon, the disaster in the Guérin canefields, and various earthslides and fissures which had disrupted the countryside and turned St. Pierre into a refugee center. Pelée had claimed over two hundred lives, a not inconsiderable number in a small colony, and Mouttet was content to state simply that "an eruption of Mount Pelée has taken place!"

While a political hack like Mouttet was adept at manipulating the terms of government service to suit his personal convenience, he was far too shrewd a veteran to make any obvious moves that might be grounds for a charge of dereliction of duty. And yet one of the primary duties of a Colonial Governor was to keep the Ministry of Colonies completely informed. Mouttet was required to submit accurate and detailed reports on many facets of colonial life, so what could have motivated him to sit down alone this Sunday night and virtually invent a situation report?

We can only conclude that Louis Mouttet had so lost touch with reality that he really *believed* that Pelée was not a threat. He had sufficiently taken leave of his senses to believe that at least two hundred deaths, numerous casualties, and widespread alarm and destruction, of all of which he had personal knowledge, could really indicate that "the eruption appears to be on the wane."

MONDAY

May 5, 1902

CHAPTER ELEVEN

The Pavement Slaughterhouse

ALONG THE WATERFRONT a small army of women were assembling in the morning sunlight. The semi-nakedness of their figures, the powerful shapeliness of their torsos, the gathering and folding and falling of light robes oscillating over swaying hips, the sculptured symmetry of bare feet scuffing the dust—these were *les porteuses*. Descended from slave stock out of Africa, their ethnological characteristics had become blurred in two hundred years by all those indefinite powers that shape the mold of races— blending of blood, habits, soil, and sun. Now they were a race apart: light skinned, firmly built human thoroughbreds epitomizing grace, force, and economy of strength.

Not one of the women was fat or old. Theirs was a world of supple limbs and youth; at forty, youth and health spent, they would seek other work, unable to compete with a new generation of girls ready to tax their bodies to their utmost capacity of strength, endurance, and rapid motion. To achieve this they trained with the dedication of true professionals. At an early age, around five, they practiced carrying a *dobanne* full of water on their heads; hour after hour they would walk through the streets carrying the clay pitchers. Then, around the age of ten, they began carrying a *trait*, a heavy wooden tray with deep outward sloping sides filled with up to fifty pounds of fruit and vegetables. Finally, around sixteen—lithe, vigorous, all tendon and hard flesh—they become full-fledged *porteuses*. For four dollars a month they worked twelve hours a day, six days a week, supporting loads of up to 150 pounds in baskets on their heads.

The four hundred young girls who waited on the water front of St. Pierre were the *porteuse* coalwomen of the Compagnie Générale Transatlantique.

All day, the sweat running into their eyes, singing as they worked, they would normally empty the warehouses of Fernand Clerc, load the barrels of rum and sugar on flatbottomed tenders, and ferry them out to the ships moored in the roadstead.

They waited now for Julie Gabou to arrive. She reigned as queen over St. Pierre's one thousand *porteuses*. Twenty years old, with the build of an ebony statue, she could hoist up to 180 pounds on her head and carry it to Fort-de-France, fifteen miles away, without a pause. She was six feet tall, strength and grace united from neck to heel.

Now, smoking a *bout*—the long, thin cigar of Martinique—she walked on to the water front.

There would, she said, be no work that day as a reprisal against Dr. Guérin for making his workers break the Sabbath. One of the men who had been crushed in the fissure had been her common-law husband. Mixing grief with industrial retribution, Julie Gabou was showing that *la fille de couleur* was a force to reckon with.

Two hundred sixty-seven years had passed since the first colored women slaves—the original *porteuses*—had been brought to the island. In 1665 the *Code Noir* had given them freedom, if not complete liberty. But as miscegenation increased, a revised Black Code in 1724 forbade marriage or concubinage between the races. It appears to have had no more effect than the previous law. And by the eighteenth century the slave women had begun to exercise their proven sex appeal to obtain their wants.

To Lafcadio Hearn, the island's historian, the emergence of the slave women to power was barely credible: "Scarcely a century had elapsed since the colonization of the island, but in that time climate and civilization had transfigured the black woman. After one or two generations, the *Africaine*, reformed, refined, beautified in her descendants, transformed into the Creole Negress, commenced to exert a fascination capable of winning anything. Travelers of the eighteenth century were confounded by the luxury of dress and of jewelry displayed by swarthy beauties in St. Pierre. But the Creole Negress, or mulattress, beginning to understand her power, sought far higher favors and privileges than silken robes and necklaces of gold beads: she sought to obtain not merely liberty for herself but for her parents, brothers, sisters, even friends."

She obtained it the only way she knew: like Lysistrata she put a price on her favors. To Hearn: "So omnipotent was the charm of half-breed beauty that masters were becoming the slaves of their slaves. It was not only the Creole Negress who had appeared to play a part in this strange drama which was the triumph of nature over interest and judgment; her daughters, far more beautiful, had grown up to aid her and to form a special class. That which only slavery could have engendered possible, began to endanger the integrity of slavery itself." So emancipation advanced. Nature, in the guise of the slave girls, had mocked at slave codes.

Now, born free, their descendants stood on the water front and exercised their right to collective bargaining.

It was the first time the women of Martinique had gone on strike. Typically, the coalwomen announced the fact by singing. With Julie Gabou at their head, they marched off the water front, chanting the words of a Creole folk song that told of their fight for freedom.

Out in the harbor, the *Pouyer-Quertier* was weighing anchor to resume her search for the broken telegraph cable. On the bridge, Captain Jules Thirion paused from watching the preparations for sailing as the sound of the singing coalwomen drifted across the water along with the tropical odors of St. Pierre awakening. To him the singing sounded "like a celebration," and looking up toward the mass of Pelée—high, towering above the town in cold silence—he wondered whether the celebration was somehow connected with the volcano. Since the black cloud had brought a momentary fear scudding over his ship the day before, Pelée had given no indication of behaving unusually. The cauliflower-like clouds had been replaced by no more than wisps of cotton around its summit.

In his own mind he was satisfied that whatever threat St. Pierre might have faced from the volcano, it was, for the time being anyway, over. A mile offshore from the water front, he could not distinguish the ash, the cracked walls of the mulatto quarter, the remains of the broken Pont Basin bridge, the disrupted telephone cables, the Roxelane River still in flood, the refugees from Le Prêcheur, the survivors from Ajoupa-Bouillon, or a hundred other things which would have told him that the situation in the town was still critical.

The threat of contamination from the animal carcasses and the occasional body which had disturbed him the day before had gone, carried out to sea. Even the ash on the surface of the water had visibly thinned; it was barely coating the chain-cable as it rumbled up through the hawseholes in an iron torrent.

Below him the deck started to shudder, and at the stern there was a whirling and whispering of water being whisked into a foaming stream as the propeller bit hard.

Steadily and swiftly, the land swung slowly round from the bridge, and St. Pierre and the *pitons* and *mornes* behind it veered and changed place and began to float away. Only the massive might of Pelée remained constant as the cable ship headed out to sea.

In the short while St. Pierre had become indistinct; its face had merged into the hinterland. Then the island itself had become a silhouette—similar to the outline it had offered Christopher Columbus from the deck of his caravel on another Monday morning four hundred years previously.

Satisfied that the plotting of the ocean floor had been properly resumed, Captain Thirion left the bridge for breakfast.

Dr. Eugène Guérin was breakfasting alone in the dining room of his villa. From outside came the familiar sounds of the sugar factory coming to life: the rustle of bundled sheaths of cane being fed to conveyor belts which carried it in a never ending stream from the wooden-

The Pavement Slaughterhouse 131

wheeled carts into the factory itself, and the steady rumble of the dynamo which drove the machinery that chopped, shredded, and crushed the cane into juice.

He had just finished breakfast when the sound of screaming came from the kitchen. Moments later Mary Goodchild, the English nurse and governess, raced through the dining room slapping at her body and shouting: "*Fourmi-fou! fourmi-fou!*"

The cry was taken up in other parts of the villa, and from the sugar factory. Dr. Guérin rushed outside to see his head overseer, Joseph du Quesne, organizing the workers to fight the plague of ants and centipedes which had swarmed in from the banks of the Blanche River.

They had been driven farther and farther down the slopes of Pelée by the ash falls; the *fourmi-fou*, speckled, yellowish creatures whose bite stings like a red-hot needle, and the *bête-à-mille-pattes,* many of them a foot long, black, with mouths capable of biting through shoe leater.

Thousand upon thousand were swarming into the factory yard, creating near panic among the workers and bringing terror to the horses as the ants and centipedes climbed up their fetlocks.

Behind him, Dr. Guérin heard the rattle of shutters closing as the villa prepared to resist the invaders. Undaunted, the ants and centipedes had started to climb the villa walls. Workmen pounded them with canes, splattering the walls with blood. From inside the villa came screams and shouts as the working women fought off the insects.

Already the centipedes had moved upstairs, lurking in bedrooms and parlors, nestling in bedding and night clothes. Mary Goodchild was hunting a group of them when she became aware of a prickling of feet on the back of her neck: a centipede had climbed up the back of her dress. With a fresh scream she raced for help, unaware that in her panic the centipede had been dislodged.

Outside, the battle was in full progress. Barrels of oil, used for lubricating the machinery, had been emptied into the yard. File after file of ants and centipedes were trapped in the thick black morass. A bucket brigade, led by Eugène Guérin, was pouring buckets of water over the terrified horses to drown the insects. Another group, seemingly oblivious of the bites, were systematically beating their way through the sugar factory, crushing the insects under damp sacks.

At the wall, over which new hoards still swarmed, Dr. Guérin and Joseph du Quesne had fired some oil-covered planking in the path of the insects. Other workmen were busy coating the wall with oil to repel the droves.

The plague which had invaded the Guérin factory was not uncommon. In 1851, when the volcano had gushed ash, the creatures had swept into the coastal areas, creating great havoc. In places they had destroyed whole plantations. Babies in their beds had been eaten alive. Immense balls of living ants had rolled into the sea in the north of the island, drifted down the coast, and come ashore at St. Pierre and Fort-de-France. It had taken months to route the insects and drive them back to the higher slopes. Behind them they had left a trail of dead domestic animals and a whole range of new superstitions to fill the mulatto minds. When cornered, went one legend, a centipede would leap at its attacker's face, and could not be dislodged without cutting it to pieces; its footmarks would have left certain living and ineffaceable marks upon the skin, a condition known as *ca ka ba ou lota*.

No one had actually seen these marks, but stories about them had been handed down from generation to generation. Another legend claimed that to kill a centipede meant a gift of money would soon fall due; even to dream of killing one brought good luck. Over the years the saying had become so universal that normally intelligent people subscribed to it. In the process the centipede had taken on the attributes of a human enemy.

So this morning the normally reserved, autocratic Dr. Guérin led his workmen in a litany of abuse against the invaders.

With every blow he shouted: "Out of my house or be damned! Get out of my house, you murderers!"

From inside the villa the cry was repeated as centipede after centipede fell under the flailing pans and flatirons of the household. Retreating, under attack, both the centipedes and the ants displayed cunning, fighting to the last bite.

The Pavement Slaughterhouse

In Eugène Guérin's house, his brother Joseph and the housemaids were engaged in a running battle with a score of centipedes, each a foot long, with pink bellies and violet heads. The creatures, their legs moving in panic causing their bodies to lengthen and shorten, were darting from room to room with stunning speed.

Holding an iron cauldron of boiling water between them, the maids tipped its contents over the creatures.

There was a crackling sound as the steaming water cascaded over the centipedes. They curled up in protective cocoon shapes. One or two of them made a last desperate sortie. Centipedes on a frontal attack are an unnerving sight, but the maids kept their nerve. They hurled the cauldron itself down on the heads of the creatures, crushing them to death.

Louis Labatut, a stoker employed at the refinery, badly bitten by ants and centipedes, had come to his own decision: no matter what the inducements, threats, or fears of future unemployment, he would not stay a moment longer in the factory. Without waiting to collect even his bottle of *tafia*, the staple drink of sugar cane workers all over the island, he ran from the yard toward St. Pierre.

Outside, the invasion had been contained. But soon there would come a new, far more terrifying menace from the Blanche River.

In his Presbytery in Morne Rouge, Curé Mary pondered at the miracle that had happened. At midmorning a group of coalwomen from St. Pierre had arrived in the village, carrying essential foodstuffs. They had been sent on the instructions of the "Action Committee." Later, *Les Colonies* would make political capital from this gesture, pointing out that the Governor had intervened to relieve the suffering of the mountain village. But now, as the coalwomen distributed the dried fish, vegetables, and white beans among his parishioners, Curé Mary could only think that God had intervened in answer to the nightlong vigil of prayer he had undertaken before the church shrine, where in the past other miracles had been prayed for and received.

For this devout Catholic the only way to acknowledge such intervention was by further prayers.

He left the Presbytery. Outside in the single, straggling street, the villagers were taking food back to their wooden cottages built into the hillside. Normally the cottages were screened by banana trees, Indian reeds, and wild roses. But the ash had killed the vegetation. Only the four palm trees grouped around the entrance to the church seemed to have survived the fallout.

Out in the street Curé Mary glanced, as he always did nowadays, in the direction of Pelée. Its head, hazy in the sunlight, towered over the shoulders of intervening *mornes*. It appeared to be totally lifeless, and Curé Mary, in his own words "ever hopeful of further bounty," wondered whether God had also answered his prayers about the volcano.

In many ways the island was outwardly more Catholic than the Vatican itself. Every road, every path, was flanked with shrines, statues of saints, or immense crucifixes. Every town or village home—whether it was an imposing stone residence, wooden cottage, or palm-thatched *ajoupa*—held a *chapelle:* an ornate wall fixture for holding crosses, vases, lamps, and wax tapers. Statues were placed in windows, above doorways, in dormers.

Curé Mary's own room in the Presbytery resembled a religious museum. Its *chapelle* held eight Virgins, a St. Joseph, a St. John, a crucifix, and a host of minute hearts or crosses, each having some special religious significance to the priest. Like so many of the religious emblems which littered the island, Curé Mary's display of crosses and statues had little to commend them as art; some of them bordered on the grotesque, jarring the esthetic senses.

To the outside world these silent populations of plaster and wood and stone were visible evidence that the Church remained rich and prosperous in Martinique. Roman Catholicism was one of the elements of the common culture which theoretically bound the island's Creole society. In St. Pierre and Fort-de-France, the Church did, in fact, still hold a place in the affection of the population. But in the country, where the influences of voodoo were powerful, Christianity was locked in battle with the pagan beliefs of Africa. In many parts of Martinique, the *quimboiseur*, or wizard, already wielded more authority than the priest, exercised more terror than the magistrate, commanded more confidence than the physician. The images and crucifixes still received respect, but there were many, like Curé Mary, who felt that this respect was inspired by a feeling that was purely fetishistic.

The priest had been on the island forty years. In that time he had seen the French Government reach out from Paris to expel various religious orders and establish lay schools where the teaching was largely noted for its aggressive antagonism to Catholic ideas. But in spite of it all, the Church had survived, though its once total hold on the maintenance of social order had been weakened.

Now, at the start of a week in which Martinique would face its gravest crisis, the Church was effectively led by a prelate who held little influence with the Governor or the lay administration. Prayer, as the Vicar-General and Curé Mary were soon to find out, would not be enough when the crisis came.

At the southern end of St. Pierre's water front, near the Savane du Fort, a topic of burning interest occupied the women who sat there. There was news of the ripening romance between René Cottrell and Colette de Jaunville. It had been brought to them by Marguerite, the old *blanchisseuse* at the villa where René was staying with his uncle and aunt. For sixty years Marguerite had brought the household linen down to be washed in this quiet tributary of the Roxelane River and afterwards left it to bleach upon the huge boulders of porphyry and prismatic basalt along its banks. With the other washerwomen, their faces hidden beneath immense straw hats, she had stood knee-deep in the water rubbing her wash.

But the ash falls had turned the water into a thick gray paste. For five days now the washerwomen had come to the stream, only to find that each time the water was grayer and dirtier. Their work impossible, they sat on the rocks in the hot sun, exchanging animated gossip. The news that Marguerite brought was exciting. On Thursday night, Ascension Day just three days off, the de Jaunvilles were holding a big party, and it was strongly rumored that the climax of the evening would be the announcement of the engagement of Master René and Miss Colette.

The news was greeted with shouts of excitement; in the world of *les blanchisseuses*, legal marriages were rare.

Yet for all its moral freedom—where industrious, affectionate women were happy to make casual liaisons with men who were often brutal to them—the world of the washerwomen was governed by rigid rules. The most professional occupied the upper end of the stream; the casual house servants who washed linen twice a month as part of their family service were relegated far downstream. To Lafcadio Hearn the washerwomen belonged to "some earlier civilization. The majority are black, for it requires a skin insensible to sun as well as the toughest of constitutions to be a *blanchisseuse*. She is the hardest worker among the whole population, and during the greater part of her labour she is working in the sun, standing up to her knees in water that descends quite cold from the mountain peaks. Her labour makes her perspire profusely and she can never venture to cool herself by further immersion without serious danger of pleurisy. The trade is said to kill many who continue at it too long." The professionals, apart from occupying the upper end of the stream, have a speed and style of their own, characterized by whipping the washed linen on the rocks with a satisfying noise. To Hearn "it is not a sharp smacking noise, as the name might seem to imply, but a heavy, hollow sound exactly like that of an axe splitting dry timber." To achieve it could take years of practice. Theirs was a dull, backbreaking way of life, and the last few days had been a welcome respite.

The news of an engagement party meant one thing, above all, to the washerwomen. They would have to find some place to launder the pantaloons and *douillettes* of their masters and mistresses before Thursday.

In the mulatto quarter, the news of the coalwomen on strike had caused great excitement. The streets, already full with refugees, had become completely blocked as people had stopped to discuss the strike. Several of them asked Father Alte Roche whether the Church would condone such industrial action; he shrugged off the question.

The Pavement Slaughterhouse 137

He was in a hurry to reach Mount Verte, from which he would spend several hours observing Pelée.

He was halfway down the Place Crocquet when he heard screams coming from ahead. Down the street came a throng of running people, and behind them came the cause of their terror: a maddened bull.

The animal, along with other cattle, had been driven into the town by refugee farmers and penned in the wooden stalls of the mulatto market. Crazed for food and water, the bull had broken out of its stall and gone on the rampage. His freedom was short-lived. Father Roche heard a "volley of shots," followed by bellows of pain, and a cheer from the fleeing crowd. A group of soldiers, on patrol to avoid looting, had killed the bull. Now, the danger over, a change came over the crowd. They turned and "almost leaped upon the carcass, their knives hacking out the juiciest pieces."

The pavement had become a slaughterhouse, with men and women carving up the animal in an astonishingly short time. In spite of the food carts, St. Pierre was still a hungry town.

To Clara and Thomas Prentiss, on their way to the cable office, the sight was "the most horrific we have ever seen. It was almost inhuman the way they cut up the beast." Sickened, they hurried home, forgetting the purpose of their mission.

CHAPTER TWELVE

A Mountain on the March

FROM HIS CELL, the world of Auguste Ciparis had suddenly narrowed down to a plank of wood, six feet long and a foot wide, standing on blocks two feet off the ground. The raised plank, like a bier, stood in the center of the prison yard. All morning the ringleaders of yesterday's riot had been brought to it, stripped naked, and laid face down on the plank. Their hands had been tightly tied under the plank; their feet were anchored in place by iron clamps built into the wood. Then, a guard, a massive man wielding the *bastonnade*, had administered the lashes. After each prisoner had received his punishment, he was untied and unclamped and rolled off the "justice bench." Then two trusties picked him up and carried his bleeding body, raw from shoulders to buttocks, to the solitary confinement cells.

Twice the trusties had carried a prisoner to a small building on the far side of the yard. Auguste Ciparis felt he knew that building almost as well as he knew his own cell. It was the mortuary. On Thursday morning he would be carried there by trusties after he had been cut down from the gallows. Later he would be placed in a coffin, covered with quicklime, and buried.

Yet this morning something had happened to rekindle his hopes of reprieve. A couple of trusties had come to his cell on the pretext of cleaning it, then clouted him with two short clubs they had concealed inside their shirts. As they left, the guard had explained to Auguste Ciparis that the beating was "to make up for the trouble you caused over the gallows." Later, he caught a few snatched words from two of the guards on duty at the "justice bench." They had said something about getting the scaffold dismantled. The words fired his optimism.

In his office in the Governor's Residency in Fort-de-France, Louis Mouttet was also feeling optimistic. It is hard to see how he could seriously justify his optimism. While the report from St. Pierre did indicate some improvement and the immediate danger of starvation had been averted, the situation remained critical. Pelée was still alive; the rivers, especially the Roxelane, were still flooding; the electricity supply had not been completely restored to the town; and the fears of the population had by no means been quelled.

It is impossible to understand how Louis Mouttet, if he had full grasp of the situation, could justify this morning's entry in his official Governor's diary. He wrote that the fading glow at Pelée's throat could be interpreted "as evidence that the crater was behaving as it did in 1851," since the volcano had shown no active signs of life for almost a day.

He waited now for one other piece of evidence, the report of his Commission of Inquiry.

The Commission was making slow progress. In the green gloom of the Jardin des Plantes, its members carefully inspected and tabulated the evidence of destruction that Professor Landes pointed out.

Entering the Botanical Gardens, they had found themselves in a green twilight filtered through the interlocking foliage overhead. Palms and tree ferns were hung with festoons of giant lianas and tapering cords of creepers. Parasitical vines with stems as thick as cables coiled about many trees until it became impossible to see their bark, and each tree seemed a pillar of leaves.

Those pillars, like all else in the Jardin des Plantes, were dead, killed by the ash. It lay everywhere—on the stonework, the grottos, bridges, basins, terraces, steps—hiding the moss and the dark stains of time. Now the Jardin des Plantes looked more like a moonscape than anything else.

But even before the ash had fallen, the Jardin des Plantes was little more than a wreck of what it once had been. Since the collapse of the French Empire, it had been shamefully neglected and abused. Despite this neglect, Lafcadio Hearn had described the gardens as suggesting a spirit of art as old as Versailles. They seemed from another century.

Professor Landes could be forgiven for believing that nature, through Pelée, had decided to complete the destruction that man had started. It would, however, be hard to justify the importance he would insist on giving to the destruction in the Jardin des Plantes at the expense of the damage caused by Pelée to the rest of the town. One-third of the eighteen pages of the Commission of Inquiry's report on the safety of St. Pierre would be devoted to the botanical gardens.

In the mulatto quarter, nature had unleashed a new attack on the town; a swarm of ferdelances, deadliest of all the snakes that roamed the island, had slithered in from the interior, driven, like the ants and centipedes that attacked the Guérin household earlier, by a search for food. In minutes they had struck terror into hundreds of people.

Six feet or more in length, its body colored yellowish-brown on top and pink on the belly, the ferdelance attacks with bewildering speed. Within the span of a few heartbeats, the pierced flesh of its victim chills and becomes soft and swollen. Spots appear on the skin, and death comes with an icy coldness creeping through the bloodstream. Later there is a terrifying necrosis of the tissues, the flesh appearing to hang from the bone, and the colors of its putrefaction resembling the hues of vegetable decay: ghastly grays, pinks, and yellows.

Fanning out, the snakes drove deep into the mulatto quarter, attacking anyone in sight, writhing, stretching, springing, striking—all in one blinding movement. Within thirty minutes they had claimed as many victims, a number of them children. Horses, dogs, hens, and other animals fell before them, too. A sow, trying to protect her litter, was attacked by three of the giant snakes, who afterward destroyed the litter in a series of tongue flicks.

The flight from the mulatto quarter would have become a full scale stampede had it not been for the troops summoned by Roger Fouché, mayor of St. Pierre. Pushing through the mulattoes, the soldiers advanced on the houses, shooting the snakes on sight.

The noise of musketfire had a curiously quieting effect on the crowd. They turned to watch the battle.

Soon, the muskets were aided by another weapon—the huge cats that strayed through the streets of St. Pierre. Attracted by the excitement, the animals had descended upon the mulatto quarter.

Mayor Fouché, looking down on the quarter from the balcony of his office, witnessed, as he told Father Alte Roche later in the day, "one of the most remarkable battles I have ever seen. The cats, whether through hunger or bravery, advanced boldly on the serpents until they were just outside striking range. Then the cats would feint, teasing, startling, or trying to draw the serpents to strike first. And when they struck, heads hissing from coiled bodies, the cats pawed them aside, claws mangling the heads of the ferdelances. Again the snakes would strike—and again the cats would sweep the heads aside, inflicting further damage. Blinded, scaled skin deeply torn, eye sockets ruptured, the ferdelances were stunned. Then, the cats leapt on them, nailing their heads to the ground while keen white teeth severed the vertebrae of the serpents."

In an hour over one hundred snakes had been either shot or bitten to death. By then fifty people and five times that number of animals had died from their venom.

The boar hunting party which had left St. Pierre the previous evening was back on familiar ground, moving on the periphery of the belt of forest that from St. Pierre had looked like a band of moss belting the volcano.

All morning they had moved across the forest edge,

hunting without success. Nothing had moved among the lofty tree trunks, lianas, or creepers. They decided to penetrate deeper into the *grand-bois*.

Higher and higher they climbed, above them always the same faint green light, below them a stairway of slippery roots.

Tropical trees do not root deeply; instead they send out great far-reaching webs of roots that interknot with other webs, and these in turn join up with still others. Between the roots, creepers and a nameless multitude of tough shrubs grow, as well as mosses, grasses, and fern. The result is that square mile after square mile of forest is locked and bound together into one woven mass, solid enough to resist the force of a hurricane.

There is only one way to cut through this mass and that is with the machete. Taking turns, the hunters, all trained cutlassers, cut their way through the forest, severing with a single blow roots and vines as thick as a man's arm. They worked without apparent difficulty, cutting horizontally so that the severed roots would not be sharp and dangerous.

They had been cutlassing their way through the forest for some time when from above and to the north they heard a dull roar.

Ahead they could see that the trees were thinning, as the *grand-bois* petered out around the stone collar of Pelée's neck. Above, too, there were glimpses of sky. Cutlassing and the need for quiet stalking were forgotten as the roar increased. Racing across the ground, the hunters gained the edge of the forest in minutes and stood upon the ridge line.

Below them to the south and west they could see the shore around St. Pierre, the town itself nestling in a curve of the coast. Over the crest of forest they had just come through, they could see the Guérin refinery. A few hundred yards to the north of where they stood, hidden by some of the fantastic rock masses that scarred Pelée's neck, was the source of the Blanche River. The roaring sound came from that direction.

Then, as the hunters stood there, the ground seemed

A Mountain on the March

to heave up ahead of them, as a giant wall of mud, nearly one hundred feet high, rose into the air, hung there for a brief moment, and then flopped down on the jungle and began to descend the slope with ever increasing speed toward the coast below. Only the Guérin factory barred it from the sea.

Louis Labatut, the stoker who had fled from the Guérin sugar factory, had reached a point halfway to St. Pierre when he heard the roar. He looked behind him. For a moment he could see nothing. Then out of the *grandbois*, two thousand feet up the side of Pelée, he saw a "great brown mass" hurtling down toward the refinery. It was about four miles away.

To Labatut, the mass appeared to be "a mountain on the move," traveling down the bed of the Blanche River. In fact it had overflowed the river, and by the time it had reached the edge of the forest it was over two hundred yards across and about a hundred feet high: a wall of millions of tons of mud, lava, boiling gases, boulders, and jungle debris traveling as fast as an express train. It continued straight on course for the sugar refinery.

In the factory yard, after the ants and the centipedes had been driven off, Dr. Guérin had faced a new problem: the men had again refused to resume work. For an hour he had reasoned with them. He was on the verge of success when Julie Gabou, the newly widowed queen of *les porteuses* in St. Pierre, arrived.

To Joseph du Quesne, head overseer, "she was like a devil appearing. She told the workers that they had been punished already for breaking the Sabbath, and that they would suffer more troubles if they worked for a man who seemed to treat their lives so lightly."

Dr. Guérin, outraged at this challenge to his authority, threatened Julie. This provoked anger from his own workers. Several of them returned to their shacks to stage a sit-in strike.

Then into the yard came the superintendent from the upriver Isnard plantation, Louis Clemencin. He brought news that his plantation workers had been sent to the safety of St. Pierre, and that farther down the coast his master, Mark Saint-Marie Moregut, was waiting with the family yacht to evacuate the Guérin household.

"Clemencin came hurrying up to Dr. Guérin, saying that his master felt they should leave at once. It turned out that the ants and centipedes had attacked their plantation, too, and Monsieur Moregut feared further happenings," recalls Joseph du Quesne.

At first Dr. Guérin refused to leave his estate. Then his wife Josephine stepped in.

"Her influence was considerable—and this time she used it to perfect effect," Joseph du Quesne remembers. "She persuaded her husband to evacuate the estate, and to take advantage of the offer of passage to St. Pierre."

Hardly had the decision been taken than the wall of mud was spied racing down the lower slope of Pelée.

Incredibly, at first nobody moved. Then Louis Clemencin shouted, according to *Les Colonies* later: "Hurry! Hurry! Get away quickly, or you are lost! An avalanche is rolling this way. Run!"

Joseph du Quesne, standing near Clemencin, recalled, also in *Les Colonies*, "that nobody said anything until I shouted for people to run. By then Clemencin was running out of the yard. I shouted again to the work people to run, and not without trouble I succeeded in getting some to move. Then I fled."

In the yard behind him, the Guérin household stood transfixed—Dr. Guérin; Josephine, his wife; his son Eugène and his wife; his daughter Sarah and his younger son Joseph; Mary Goodchild, the maids, and the cook.

Joseph du Quesne had run several hundred yards clear of the refinery when he looked back: "The Guérins were running too, now, and from above the factory I saw the wall of mud approaching. It was high enough to blot out all behind it, and very wide."

Later, it was estimated that the mud wall was, in fact, about 120 feet high at this point and about a quarter of a mile wide.

For a split second it hung over the factory and then fell with a splintering crash.

To Joseph du Quesne, the sight was "impossible to

put into mere words." He stood on a hillside and watched as the flow of mud obliterated forever the refinery where he had worked for forty years.

"It was as if it had never been there. It was as if the people I had seen in the yard, whom I knew and loved, had never been there."

The wall of mud, its speed slowed under the impact of absorbing the Guérin refinery, was rolling on down to the sea, spreading even wider as it moved.

Joseph du Quesne did not see what further damage it inflicted. With his face in his hands, he wept uncontrollably.

Louis Labatut had ample time to contemplate the disaster unfolding before his eyes. Later he was to tell the readers of *Les Colonies:* "At the outer edges of the mud flow, I could see people struggling to free themselves. One of them was Julie Gabou, the coalwoman. Then moments after the factory had been drowned, which was a little after twelve o'clock, a boiling waterspout burst from the mountain, leaping all obstacles in gigantic bounds, flowing over the already vanished works and killing all those still struggling in the mud. Then I saw that the mud and the waterspout were no more than a terrible prologue. Behind them a torrent of water laden with rocks and earth came pouring down the side of Pelée to raze the whole region and to form a mud plain that extends from the sea to the *grand-bois*. The mud destroyed the yacht of the Moregut family as it rode at anchor, waiting in vain for the household of Dr. Guérin to appear. Where there had been sea, there was now nothing but mud."

As he stood there, he saw three figures coming toward him. They were Joseph du Quesne, Louis Clemencin, and, incredibly, Dr. Guérin. He had not run with the rest of his family, and by a miracle, the mud had not reached the spot where he stood behind his son's house.

These three were the only survivors of the catastrophe. So stunned was he by what had happened, Dr. Guérin was nearly unconscious. In silence the other men carried him down the road to St. Pierre.

Behind, beneath the mud, they left 159 bodies. Most of them would remain forever where they had been choked to death. Only the tip of the chimney stack remained above ground to mark their burial place.

CHAPTER THIRTEEN

The Great Wave

FROM A POINT near the top of Mount Verte, Father Alte Roche had watched the avalanche of mud bury his theory that Pelée presented no threat. In a few moments a wide brown swath had stained the patchwork of swamps, streams, and jungle below him. It had obliterated the Blanche River, the sugar cane fields on either side of it, and the defenses erected upstream to prevent flooding, as well as the refinery itself. All that remained visible of the building was the tip of its eighty-foot chimney stack, held up by two of its original eight cables. It was still smoking.

The moments of devastation were clearly imprinted on the Jesuit's mind: "Hardly had the midday hour passed on this Monday when the gates of the volcano were drawn and a flood of boiling mud was sent hurling down the mountainside to be flung from it into the sea. In three minutes it had covered its last three miles to the ocean, and within that time it had left nothing visible of the Guérin refinery but the chimney—a post projecting from a desert of black boiling and seething mud. The factory had stood as a symbol for what it represented through long years of toil and conquest. Now it had disappeared as if the hand of the Devil had covered it."

From his vantage point, looking out across the *grand-bois* to the source of the Blanche River, Father Roche could clearly see where the mud had begun its journey of destruction: it was still pouring out—"like pus from an infected wound"—from a huge hole in the side of the volcano. Blocks of rock estimated by Father Roche to "weigh as much as fifty tons" had burst out of the side of the volcano.

Looking down on the disaster area, Father Roche estimated the "depth of the mud in some parts of its flow was probably not less than 150 feet. By the time it had reached the coastline, it had spread to cover a distance of nearly a mile. Nothing moved in the area."

But now a new terror was building up. The mud was driving the sea farther and farther offshore. The water, fighting to resist the huge pressures, was rearing higher and higher into the air.

On the bridge of the *Pouyer-Quertier,* by now three miles offshore, Captain Jules Thirion could not believe what he saw through his telescope. The coast was disappearing behind a wall of water.

From the deck hands came shouts of "Tidal wave! Tidal wave!"

But in all his experience Captain Thirion had never known a tidal wave that rose out of the surf close to the shore. Though the difference would matter little to its victims, it was in fact a shock wave and not related to the tides.

At that moment smoke, thick and gray, started to pour from Pelée's throat. It hung in the still air, then drifted down on the countryside. From her bedroom window Suzette Lavenière watched the billowing smoke roll over the countryside. After leaving the hospital with Father Roche, she had been escorted home by a group of workers from the estate. Safely inside the villa, she had stepped out of the nurse's robe, bathed, and gone to bed. She had slept for nearly twenty hours. Now, refreshed by the rest, she had started to take hold of the reins of the plantation. At twenty-four she had stepped into a role that a week before she would not have thought possible—mistress of all she could see from her bedroom window. But even now Pelée seemed determined to flaunt its mastery over her; the smoke was rapidly blotting out her land under a gray haze.

The Great Wave

From her balcony Véronique Clerc thought the sea "standing on end" just north of St. Pierre was a trick of the light. She was still wondering what had caused this illusion when the dust from Pelée drifted across her view, blocking out the town and the sea beyond.

The Governor's Commission of Inquiry had finished. After leaving the Jardin des Plantes, an hour had sufficed for its members to have conducted what they were to call "the most detailed examination of the safety of St. Pierre." Now, satisfied that they had seen all they needed to, the Commission members were on the way back to Fort-de-France. They were in a hurry. Louis Mouttet had sent them an invitation for lunch, and his lunches were famed throughout the island. They had reached the summit of the hill overlooking St. Pierre when the smoke started belching from Pelée. Professor Landes, having firmly established himself as the Commission's expert witness on the volcano's behavior, was of the opinion, according to Lieutenant Colonel Jules Gerbault, President of the Commission, that the clouds "while impressive presented no more a threat than the previous emissions."

Again Professor Landes had seriously misjudged the situation.

If he, or the other members of the Commission, had looked back toward St. Pierre and the roadstead beyond, they would have seen convincing evidence against their optimism.

Out in the roadstead Captain Marino Leboffe of the *Orsolina* for some little while had been watching the stirring of the sea as it stroked the side of his ship and moved the other ships riding at anchor nearby. It was these movements which had disturbed Captain Leboffe. The sea, like the earth, is ever changing. Most of these changes the Italian recognized, but for some time now there had been a scarcely detectable surge sweeping steadily in to the water front.

With the smoke from Pelée starting to drift over the town, Captain Leboffe decided to weigh anchor. In justification he was to say later: "I am a sailor. I knew there was something wrong. I could not pinpoint it, but I just knew." Shortly afterward, Captain Sequin of the *Topaz* had also weighed anchor for reasons that he too would later have difficulty in pinpointing exactly.

From the bridge of the *Pouyer-Quertier*, Captain Jules Thirion now had no doubts as to the gravity of the situation. He had not taken his eyes off the wave as it traveled, almost broadside to the coast, a wall of water weighing billions of tons, moving majestically toward St. Pierre.

He ordered the helmsman to keep pace with it. But in minutes the smoke rolling down from Pelée had blurred the view. The last glimpse Captain Thirion had of the monster wave was when it entered the roadstead of St. Pierre.

Fernand Clerc was probably the first person in St. Pierre to spot the wave. He had gone to the water front to find out if anybody had news of Julie Gabou, who now lay dead beneath the mud not far from where her common-law husband had been buried on the Guérin estate. The water front was almost deserted, the striking coalwomen had dispersed to the mulatto quarter, and other workers, unable to work because of the strike, had drifted away. Usually a scene of bustling activity, there was now an air of neglect about the water front.

He had watched with surprise as the *Topaz*, the *Orsolina*, and several other ships had weighed anchor and moved well down the roadstead. With no more than passing interest he had observed a small wave slapping against the water front. It had been followed by another, slightly larger than the first. Then another one had lapped against the timbers of the piers. It was then that he had noticed the change. This third wave had a distinct runoff, sweeping well back into the roadstead.

Then, in the distance, he saw the wall of water coming into the roadstead. In its progress down the coast, it had lost much of its height and speed. But to Fernand Clerc it still looked immense: "nearly fifty feet in height, making a noise like the hissing of a million snakes."

At that moment the ash fall from Pelée obliterated the view.

But Fernand Clerc had seen enough. Turning, he fled into the town shouting warnings.

Captain Leboffe, safely anchored clear of the great wave, watched its progress. He had never seen a natural force more menacing or more inexorable. Throughout the ship there was "the beginning of a general kind of fear— the fear man has when he sees nature with her mask torn away."

Ahead of the wave rolled rocks that had been blasted out of Pelée's side and shot down its slopes on the crest of the mud-flow; they were now being rolled along the ocean floor by the wave.

Captain Leboffe watched as the wave hit two of the Italian barks moored in the roadstead. Their moorings were uprooted, and the ships lifted high up the churning wall of water and borne on its crest toward the shore.

In St. Pierre the approaching wave caused widespread panic. From the balcony above his office, Andréus Hurard, the editor of *Les Colonies,* watched, "through a thick volume of smoke, the crowds running to this side and that in extraordinary agitation. The women especially seemed to have escaped from a mad-house. A human flood poured up from the depths of the water front. It was a flight for safety without knowing where to turn. Shop-girls were fleeing with bundles, one with a corset, another with a pair of boots that did not match; and all these people in burlesque costumes which would have caused laughter if the panic had not broken out at so tragic a moment. The whole city is on foot. Doors of shops and private houses are closing. Everyone is preparing to take refuge on the heights."

Everyone, except Andréus Hurard. He was too old to run. "If I am to die I wish to do so at my work," he wrote.

In the streets the panic was greater. Men, fleeing from the sea front, trampled over women and children.

When Emile le Cure, general manager of the English Colonial Bank, tried to remonstrate with some of the men, he was thrown back into the doors of the bank. A couple of bank clerks picked him up, and together the three men retreated inside, barricading the door behind them in a pathetic attempt to keep out the torrent which would soon pour into the town.

Alfred Descailles, the island's leading ship supplier, had been lunching with Pierre Theroset, another member of the "Action Committee," when the sounds of panic reached them. They hurried from the St. Pierre Sporting Club just in time to see a nearly exhausted Fernand Clerc go stumbling past. Behind him along the Rue Victor Hugo streamed a rabble of screaming men, women, and children. In the thick of the crowd the members of the Committee saw Robert de Saint-Cyr, the august owner of the Bank of Martinique, elbowing and shoving his way, just like everybody else.

Overhead, the dust clouds thinned as a stiff breeze blew in from the sea, whirling the ash back over the hinterland. From Pelée came a deep growling. To Professor Bordier in the grounds of the Lycée, "the sound was like the rumbling of wagons crossing a bridge, and at times like distant thunder."

As he stood looking down on the roadstead, the answer to the question he had put in his last letter to his wife came seething toward the town; this was to be the next phenomenon, and he would face it as he had faced the others—by doing nothing.

The wave now nearing the shore was watched in silence by a small crowd which had struggled to the summit of the Morne d'Orange. In it were René Cottrell and Colette de Jaunville, Clara and Thomas Prentiss. From the top of the *morne* they could clearly hear the hissing of the wave as it moved on to the water front. There was a moment when the steep front of water hung in critical inbalance. Then, in a powerful surge, it crashed down on to the water front.

René Cottrell remembers, "The barks were lifted right over the first row of low buildings and dumped with shattering crunches beyond. The great mass of water appeared to glide forward, and a line of warehouses were ripped from their foundations and splintered; the water swept this debris against other structures, and in moments the sea-front sector of St. Pierre had sunk beneath the wave."

Carts, barrels, animals, even whole buildings were swept up like bathtub toys and stacked, smashed, and drowned against buildings at the end of flooded streets. The water, almost up to the upper balconies now, was roaring up the Rue Victor Hugo. The watchers saw it breach the doors of the Colonial Bank.

Down in the town Fernand Clerc slowly collapsed at the far corner of the Place Bertin. Minutes before, it had been filled with a running mass of people. Now it was empty.

Clerc could go no further; his age and weight were against him. He turned and waited for death to sweep him up.

But when the wave finally entered the Place Bertin, it was a spent force; at no more than knee height, it washed up the sloping square.

"I lay there mesmerized. The water, try as it might, could not resist the pull that was being exerted to take it back to the sea. It started to retreat with a great sighing and hissing."

The wachers on the Morne d'Orange and other hillocks outside St. Pierre had also seen the water rolling back into the sea. taking with it as much debris as it could pry free.

When it had gone, just twenty minutes after smashing into St. Pierre, the downtown sector looked, according to *Les Colonies*, "like a place which has just been visited by the scouts of the enemy's army, the inhabitants of which have deserted before the bombardment and invasion."

Whole sections of the water front, including the warehouses of Fernand Clerc, had been pounded into splinters. The Italian barks, lifted by the receding water, became battering rams to flatten buildings; their hulks now lay on the water front, their masts stripped, their decks splintered, their crews dead or cruelly injured.

Father Alte Roche hardly believed St. Pierre could undergo more. The scene terrified him. The roaring of the volcano continued throughout the ordeal. The familiar water front now had only a few buildings still standing among the rubble left by the wave. A continuous heavy fall of ash left the town in darkness, and its inhabitants had to grope their way through the crowded streets. By the aid of flashes of lightning of almost blinding intensity, they sought, and sometimes found, their dead and missing.

By early afternoon, troops aided by *gendarmerie* had begun the task of recovering bodies and assessing the damage.

The bodies of Emile le Cure, general manager of the English Colonial Bank, and his two clerks were found under the main counter of the bank, where the water's force had swept them.

Death by drowning was also the fate of all twenty-eight children in the orphanage of St. Anne. The wave, about twelve feet high at this point, had engulfed the single-storied building, stoving in shutters and doors.

Along the lower half of the Rue Victor Hugo, the devastation was almost total. Shops on both sides of the street were crushed under the wall of water. Food, vegetables, and kitchen pots were scattered everywhere. In between lay the bodies, black, brown, and yellow—the women still wearing Madras turbans, the men with mushroom-shaped hats as large as umbrellas on their heads. In death many of them wore a suprised look; it was a look rescuers also found on the faces of a number of the *blanchisseuses* they recovered from the tributary of the Roxelane River. The washerwomen had been given no chance to run before the wave had passed over them on its way into the town.

In the mulatto quarter, the searchers recovered sixty-eight bodies of *propriets vivriers,* smallholders, who had been drowned in their tin-roofed houses. The wave had uprooted the trees and killed the goats, cows, and small Creole horses at their tether posts.

With the boulders, sea life, and mud that the water had dredged up had also come some of the bodies which in previous days had been swept out to sea. Among them was the body of Eli Victor. It had been deposited not far from the wrecked Pont Basin where, less than forty-eight hours before, his mourners had been pitched into the water and swept to their deaths.

CHAPTER FOURTEEN

A Shortage of Coffins

ON THE EDGE of the mud that had buried the Guérin refinery and the surrounding countryside, hundreds of helpless rescuers stared at the mire which bubbled and heaved like a living thing. Among these observers was Andréus Hurard, editor of *Les Colonies* and, according to René Cottrell, "a shattered man" filled with "visible misgivings for advocating in print that Pelée offered no threat."

When the huge wave had swept toward his offices, Andréus Hurard was convinced he would be drowned. But the wave, while doing considerable damage to the printing plant, had passed by several feet below his balcony. When it receded with a great slurping sound, he hurried out into the Rue Victor Hugo, to find himself caught up in "a great concourse of men, women, and children weeping, crying out, and imploring the mercy of Heaven." He had been carried along into the town, to a point where the wave's high mark was clearly visible. Beyond this line "feverish activity" was going on in many of the houses as the tenants prepared to evacuate. Hurard had watched as "handcarts transported their furniture and few belongings, piled high up in the pell-mell confusion of haste and panic. Mattresses are being carried in all directions as the whole city appears to be making their way out of the town on foot."

They didn't go far, roosting on the sides of the hillocks which lay just behind the town.

On the outskirts of St. Pierre, Hurard had met Alfred

Descailles and Pierre Theroset, who gave him news of the catastrophe at the Guérin estate.

Hurard had found a horse and hurried to the scene. In the next edition of *Les Colonies,* he recorded: "Cinder tornadoes whirled about us and choked the hardiest as we paused abruptly before a great plain, a sea of motionless mud, absolutely level, broken at intervals by little clouds of vapor, like puffs of tobacco smoke, which burst with a bubble-like sound. And the sugar works? Where are they? The glance turns to the left, toward the sea, and as far as it can reach meets nothing but the same mud plain—nothing. Nothing is left. Yes, over there is the chimney of the factory, slightly inclined, like the leaning tower of Pisa. The refinery and all their outbuildings have gone down in the lava bed. Nothing remains but that sheet-iron chimney. This is all. And here but a moment ago was a center of prosperity and activity for a world of workers, now swept either out of life or into misery and ruin. Over the scene hangs the silence of annihilation, broken only by the muffled sound of the breaking puffs of steam over the mud. The spectators cannot shake off an indescribable feeling of anguish. How many human lives have been wiped out by that sea of mud? How many fathers of families have gone forever with the rush of the avalanche? Will the exact number ever be known? Those who have been allowed to pass the line of *gendarmes* and to reach this spot stand speechless. There is nothing to say. The reality is far more terrible than anything that could have been imagined."

Louis Mouttet had some good news at last. Over lunch the Governor's Commission of Inquiry had made its report. It ran to eighteen pages. That afternoon secretaries at the Residency were busy typing out copies. Now in the cool of the evening, Louis Mouttet was reading one of them. The Commission was united in its counsel that "there is nothing in the activity of Pelée that warrants a departure from St. Pierre." The "position of the craters and of the valleys opening on the sea was such," concluded the Commission, "that the safety of St. Pierre was absolutely assured."

The Report made no reference to the flow of mud, the gigantic wave, the state of terror in St. Pierre. It made only passing reference to the disrupted telephone and cable services and the swollen Roxelane River. It did devote six pages of minute detail to the damage inflicted on the Jardin des Plantes.

A year later, a fellow professor, Angelo Heilprin, was to censure Landes for putting his name to the Report and giving it his unqualified blessing.

With the name of Martinique's most respected scientist as a signatory, Louis Mouttet felt reassured. That reassurance becomes incomprehensible in the light of a report the garrison commander in St. Pierre had telegraphed from the town earlier in the evening. It read: "Latest news. 12:55—Blanche River becomes furious torrent of muddy lava. 1:22—A moment ago a very strong eruptive thrust. The sea is rising. Shops have been engulfed. Boats beached. It is a tidal wave. The docks are smashed. Very serious situation. Panic appalling. 1:27—Sea retreats a hundred feet, returns to shore far surpassing normal level. Formation of numerous crevasses and fumaroles in valley of Blanche River. Situation very serious. Panic appalling. 1:35—A flow of lava surged down to the sea bed in less than three minutes. There are probably victims. 3:00—Guérin factory in part collapsed. All factory personnel buried under its ruins. Victims appear very numerous. The lava, arriving at the shore, gave rise to a receding action which accentuated the tide and then crashed back toward the shore, producing a gigantic wave. Appalling panic. Inhabitants of St. Pierre retreat to high ground. 3:29—The tidal wave is over. It lasted no more than a quarter of an hour. The Guérin factory no longer exists. Over a distance of two thousand feet everything has been covered by a thickness of about thirty feet of muddy lava. A vast gully has been gouged out by the passage of lava. The Blanche River no longer exists. It is completely smothered by a bed of mud about ninety feet thick."

Three times that message referred to the serious state of panic in the town and its environs. It had been sent by the garrison commander normally a phlegmatic man; he described the situation clearly as a total disaster.

A Shortage of Coffins

Yet Louis Mouttet, armed with soothing words from his Commission of Inquiry, ignored the report.

Later, it was suggested in his defense that the message never reached him because of a fault in the telegraph link between St. Pierre and Fort-de-France. No such fault was recorded by the duty officer in the Fort-de-France cable office. Moreover, as was the custom, the garrison commander had sent a copy of the message overland; it arrived in Fort-de-France late on Monday evening. Any suggestion that a member of his staff held back the message does not bear serious examination. No aide would have dared take it upon himself to suppress, or even delay, such an obviously important telegram. Therefore, there can be little doubt that the report physically reached Louis Mouttet at about the time he was studying the verdict of the Commission of Inquiry.

The Commission's verdict was a whitewash that Louis Mouttet wanted to believe. It confirmed his own delusion that there was nothing to fear; it bolstered him against people like Fernand Clerc, Thomas Prentiss, the American Counsul, and all those other people who had agitated that Pelée was a threat. The Commission's report would also quash much of the fire from the Radical propaganda guns, whose broadsides about how the volcano would calm down only when the Progressives were out of office had been having a decided effect. The report was the bromide that Louis Mouttet hungered for; behind it he could shield and if need be defend himself against further attacks.

The arrival of the garrison commander's telegram should have brought a shocked realization of the true situation. But for Louis Mouttet, pinning a childlike faith on the Commission's report, that telegram could well have been too much for his already strained mind; and his mind had totally rejected the information it was asked to digest.

A telegram of this nature would need immediate acknowledgment. None was sent. It also warranted immediate action. At the very minimum the statements contained in the telegram should have been checked out. Such a telegram would demand that the Governor should hold consulations with his staff. No meeting was called. The Ministry of Colonies would expect to be urgently informed of the developing situation. No telegram was even drafted, let alone sent, by Louis Mouttet to his superiors in Paris this Monday night. Instead he dismissed the report from the garrison commander as if he had never seen it.

The Governor was reading the Commission of Inquiry's report for a second time when an aide brought more disturbing news. The rescue workers in the mulatto quarter in St. Pierre were facing a new horror: *la Verette*. Three cases had already been reported.

In previous outbreaks whole areas of the island's black population had been decimated by the pox. With the final election under a week away, an outbreak would shatter the Radical dreams. But it is unlikely that even Louis Mouttet, conniving and politically shifty though he may have been, would be party to what in effect would be little less than mass murder for selfish political advantage. On the other hand there are those who ask what action he could have taken. Of the many courses open to him— letting Paris know, ordering all the island's available medical help to St. Pierre, to name but two—by far the most important action he could have taken was to summon help from neighboring islands. Both St. Lucia and Dominica had extensive medical facilities which could have been swiftly brought to Martinique to help control the outbreak.

Instead the Governor completely ignored *la Verette* and helped bring the population of St. Pierre just a little closer to destruction.

To Professor Bordier, tutor in history at the Lycée of St. Pierre, with the news that *la Verette* had been discovered in some of the victims dug out of the mulatto quarter came a feeling of inevitablity. The pox had found a natural breeding ground in the ash-choked sewers and waterpipes of the district. "It is as though something sluggish and viewless, dormant and deadly, had been stirred to life by the events of the past few days," he observed.

As night fell swiftly, the townspeople drifted back into St. Pierre. It was then that they heard that the outbreak of fever had spread. St. Pierre, exhausted, broken,

A Shortage of Coffins

in places on the brink of collapse, had to rally once more to meet this new threat.

Great cauldrons of tar were kindled every hundred feet along the water front, each tended by a mulatto. They were lit according to the ancient plague ritual of purifying the air. To the *Orsolina*, *Pouyer-Quertier*, and other ships that had escaped the huge wave and now rode at anchor far out in the roadstead, the fires were a clear warning that pestilence hung in the air of St. Pierre.

By early evening the list of the dead from the disasters of the day contained the names of 112 men, women, and children "positively identified." But ninety-seven others, including crew members from the wrecked barks *Sacro Cuore* and *Nord America,* were buried without anyone's knowing who they were. Their deaths brought the total number of victims that Pelée had claimed in four days to a total of 617.

In the town itself people stood well clear of the bodies, covered in quicklime, being carried to a mass grave outside the town. When the supply of coffins ran out, the corpses were wrapped in banana leaves and carried on canvas-sacking stretchers to the grave. They were still carrying the corpses away for burial when Monday ended. For all those who had prayed in the Cathedral of Saint Pierre and elsewhere that Tuesday would produce a reprieve, Pelée had a cruel disappointment. At one minute past midnight on Tuesday morning, Professor Bordier saw "long tongues of flame shooting out of the crater's neck, the like of which none of us had seen before."

TUESDAY

May 6, 1902

CHAPTER FIFTEEN

The Sympathy of the Government

IN THE EARLY HOURS of Tuesday, St. Pierre had a look of melancholia. A series of processions wound their way through the streets, chanting prayers for the dead, for the cessation of the pestilence, and for God's intervention. The more devout wore the traditional mourning garb of the island, a black robe with black foulard and turban. They crossed and recrossed the Rue Victor Hugo, weaving and interweaving, pausing before every shrine and crucifix they passed to offer prayers for Divine help. At the more important shrines many of them knelt in the ash or reached forward to kiss a cross or the feet of the Virgin. As dawn approached, the first of the processions started to make its way toward the cemetery, the Cimetière du Mouillage. Others tagged on until there seemed to be one immense line of people waiting their turn to walk slowly through the graves and shrines and then return to the town.

There, throughout the night, the searchers had dug out the last of the dead. Those with signs of *la Verette* were carried to the mass grave outside the town; the others were laid out for burial later in the day.

Sanitary conditions in the town were becoming increasingly serious, especially in the mulatto quarter. There a small house often sheltered up to thirty people. The fever swept on through an area where, as Lafcadio Hearn observed, "the poorer classes had been accustomed from birth to live as simply as animals—wearing scarcely any clothing, sleeping on bare floors, exposing themselves to all changes of weather, eating the cheapest and coarsest food." Even though living under such adverse conditions, no healthier people could normally be found; every yard had its fountain, almost everyone bathed daily in the Roxelane or the bay. But the ash had fouled the fountains, and the swollen river and the sea beyond were infected with refuse. *La Verette* had found the ideal conditions for spreading havoc.

All night long Father Roche, and the other priests of St. Pierre, administered the last rites to those beyond hope.

Now, as dawn arrived—signaled by no more than the merest lightening of a sky filled with the cinders of Pelée—the Jesuit lay on his bed exhausted, snatching a brief respite. He had just closed his eyes when "the air was filled with an extra sharp pattering of cinders, and in a moment my bed was covered with hot ash that singed the covers and pillows until they started to smoulder."

All over the town people were beating out the little fires that the latest fallout had started. St. Pierre, after five days of bombardment by Pelée, was now a gigantic tinderbox.

In the American Residency, Thomas and Clara Prentiss, exhausted from lack of sleep and the effect of the sulphur fumes—both normally suffered from respiratory problems—were discussing the worsening situation with René Cottrell. The American diplomat hoped that as a member of the Ten Families, Cottrell could exercise some influence with the Governor to have the town evacuated.

"I fear that I, or my family, would have even less success than you appear to have had," replied René Cottrell. "All we can pray for is outside intervention which will force him to evacuate the town."

"By now, the volcano dust must have been carried hundreds of miles to give the alert," said Clara Prentiss. "Surely the whole world must now be realizing our plight."

The ash *had* been carried far afield, and was indeed noticed by the world. But tragically, the observers were to make a mistake in identifying the source of the ash.

The Sympathy of the Government

Seven hundred sixty miles away, the bark *Jupiter*, out of Cape Town on a course east-southeast of Martinique, sailed at dawn through a thick cloud of lava dust. The cloud was logged without comment. Six hundred miles east of Pelée, the bark *Beechwood*, bound from Salaverry to New York, logged the falling dust as "heavy, being so far from land." Other observations were made on the steamship *Coya*, bound from Montevideo to New York; 350 miles from Martinique her log records the dust as "markedly unpleasant." Unpleasant too was the cryptic entry of the log of the *Eleanor M. Williams,* a bark from Conetable Island for New York: "At latitude 14°N, longtitude 57°W, 250 miles east of Martinique, we began sailing for eight hours through the dust." Passengers on board the Royal Mail steamer *La Plata*, out of Barbados for London—two hundred miles off the island—complained that the dust was making it impossible for them to take their usual early morning promenade on deck.

To those who asked where the ash had come from, the deck officers on the liner had a ready answer. It was probably drifting, they said, from the great Soufrière crater, on the British island of St. Vincent, which had shown signs of life for the last year.

St. Vincent lay forty-nine miles to the south of Martinique, and in the hours to come the erupting Soufrière would seriously confuse the situation. World interest, when it had finally been aroused, would center on St. Vincent, while to the north, its sister island in the Carribean chain would suffer alone.

The Soufrière was remarkably like Pelée. Eight miles wide at the base, it rose in the classic volcanic shape to a height of four thousand feet. At its summit was a crater nearly a mile across.

Near its foot lay Georgetown. The island itself, like Martinique, was thickly wooded and mountainous; its *mornes* and *pitons* bore the names of a long British occupation: Mount St. Andrew, Richmond Peak, and Dark Head; there was a Cumberland Valley and Richmond Vale; the largest beaches, wide and gentle on the windward side of the island, were called Mount Pleasant and Brighton. From all of these places, as from anywhere on the 150 square miles of volcanic rock that made up St. Vincent, the Soufrière was visible. For ninety years it had slumbered. Then in February, 1901, it had sent a series of sharp tremors through the northern end of the island. Panic had swept the Carib settlements in the area, and the natives had gone to Georgetown for safety or headed farther south to the island's principal port, Kingston.

For the next ten months reports reached the Royal Society in London that "earthquakes and noises continued. They were neither violent nor loud, but were more numerous than was usual for that part of the island; the white inhabitants regarded them with indifference or curiosity, for as earthquakes are by no means rare in St. Vincent, the repeated small shocks felt during 1901 were not regarded as necessarily the precursor of a cataclysm." The reactions of the administrations of Martinique and St. Vincent were remarkably similar in the early stages when both were content to ignore the danger at hand. But while St. Vincent had no Fernand Clerc, it did have a number of solidly British men—with names like Dunbar-Hughes, McDonald, Robertson, and Kelly—to keep track of the Soufrière's behavior.

In the latter half of April, 1902, the crater had come to a new stage of awakening; eight tremors disturbed the night of April 14th. They dislodged stones from the lava cliffs and sent them tumbling down the slopes in a series of landslides. On Monday, April 29th, around the time that Pelée was emerging from deep sleep, the Soufrière gave off three "well-marked shocks" that rocked the houses around Windsor Forest. Steadily the pressures had increased, as they had inside Pelée, until on Monday, May 5th, the Soufrière had erupted, sending great clouds of steam and dust high into the atmosphere. The settlements on its slopes had been evacuated, and neither Georgetown nor Kingston was in any danger.

But the island's Chief Constable, Captain John Calder, sent a series of messages to the Foreign Office in London on the eruption. Other messages went to Sir Robert Baxter Llewelyn, K.C.M.G., Commander-in-Chief of the Windward Islands, and Sir Frederic Mitchell Hodgson, K.C.M.G., Governor of Barbados. The cables all said the

The Sympathy of the Government 169

same thing: the situation was being carefully observed, and should assistance be needed, it would be requested at once.

By midmorning of this Tuesday, ships sailing from all the British dependencies in the Caribbean were being advised of the eruption and cautioned to steer clear, if possible, of St. Vincent.

To Captain Edward William Freeman, master of the steamer, *Roddam,* the news of the erupting Soufrière settled his homeward passage; he would bypass the island if he was fortunate enough not to have any passengers or cargo to unload there after his next stop at St. Pierre.

Captain Freeman, a tall, handsome man, had been in command of the *Roddam* for four years. He had grown to love every inch of her 289-foot, six-inch length. Built at the Withey Shipyard in West Hartlepool in 1887, the iron cargo ship was the pride of the British Scrutton Line, carrying cargo and passengers out of London to the West Indies. On this voyage she had left London on April 1st and arrived in Barbados on Sunday, May 4th. Now as she sailed from the port with the news of the eruption on St. Vincent, Captain Freeman "positively looked forward" to the extra time he could spend at St. Pierre. He was due there in two days, on Thursday, May 8th, Ascension Day.

Four hundred miles away to the north, the Quebec Line steamship *Roraima* was eleven days out of New York. Bound for Demerara, via the Windward Islands, the ship was ahead of schedule. At her present speed, seven knots, the 2,712-ton schooner would arrive at her next port of call, Dominica, in the early hours of Thursday morning.

To Captain George Muggah, standing on the bridge with his first officer, Ellery Scott, this was welcome news. It would mean that the *Roraima* could arrive at St. Pierre in time for breakfast on Thursday, May 8th. As Ellery Scott noted, this "was a chance for the passengers and cargo for Martinique to be unloaded before the heat of the day became intolerable."

It was midmorning when he observed "a cloud of

dust" drifting far down on the starboard. It was probably he thought, ash from the Soufrière of St. Vincent.

The miscalculations made by observers in the area as to origin of the dust were echoed in Paris.

The cable that Louis Mouttet had drafted late Sunday night had finally been cleared through the telegraph office at Fort-de-France on Monday morning. Because the telegram had not been given a priority clearance—probably a slip by somebody in the Governor's office which had not been corrected by the cable office—it did not arrive in Paris until late on Monday evening. By then the official working day of the Ministry of the Colonies was over. A messenger delivered the cable to the Ministry, received a receipt from the duty officer, who seemingly dismissed the message as being of little importance, understandably enough in view of Louis Mouttet's inexplicable opinion that "the eruption appears to be on the wane." There it lay all night; in the morning another duty officer studied it before passing it on to the appropriate section in the Ministry which dealt with the affairs of Martinique.

Until then it was perhaps understandable that little notice should have been taken of the Governor's cable. Hundreds of telegrams came to and left the Ministry every day; one more from an obscure colony would not create undue excitement. But with its arrival in the section responsible for the West Indies sometime on Tuesday morning, the reaction to the cable must be viewed in a different light. Those who worked in the section were theoretically supposed to have a good knowledge of the islands. Therefore a cable from one of the governors in the area it administered, stating that a volcano long dead had come to life, should have provoked at least some reaction.

The very vagueness of Mouttet's telegram laid it open to questions for further information. If "inhabitants have had to abandon their dwelling precipitately," then the island administration must have been faced with a refugee problem. How was it coping, and how did the Governor know, *for certain*, that "the eruption appears to be on the wane?" Then there were other questions that the section should have been asking: what effect was the eruption having on the island's economy and political life?

In seeking answers to these questions the section would have been performing no more than its normal duty. That minimal performance would have included one other mandatory course of action: the preparation of a situation report for the Minister. A volcano erupting on a colonial island on the eve of elections crucial to the French Administration was a situation that demanded the fullest investigation.

Yet all that happened was that the cable from Louis Mouttet was placed in a file full of previous cables from the island's administration; these contained details of census, livestock figures, exports and imports.

The Mouttet cable would remain undisturbed in this file for six vital hours before another cable, bearing the prefix "urgent priority," would arrive at the Ministry. It would cause excitement, not because of its classification, but because it misspelt the Minister's name.

Senator Amédee Knight had drafted the cable to the Minister from St. Pierre. It was a major move in the struggle he was waging for Radical political supremacy. It would also enhance his own personal position where it mattered the most, in the French Government. He saw that as the island's Senator he could gain valuable mileage for himself out of the situation by adopting the attitude of "a responsible politician appealing on behalf of his electorate for urgent assistance."

Some time on Monday night a copy of Governor Mouttet's cable had made its way into Senator Knight's hands. The most likely explanation is that it had been copied by a colored clerk in the cable office, a not infrequent occurrence, as the French Administration had long since recognized. To Senator Knight, the political capital to be made from the Mouttet cable, both for the Radical cause and for himself, was obvious.

"The Governor's cable gave no real picture of the situation as it was, or as it was likely to develop. It bore all the signs of the ineptness that the island's colonial administration was famed for," he was to state later in justifying the cable he was sending "urgent priority" to Pierre Louis Albert Decrais, Minister of the Colonies.

Senator Knight's cable read:

> VOLCANIC ERUPTION DESTROYED LIVELIHOOD POPULATION PRECHEUR EXCLUSIVELY COMPOSED SMALLHOLDERS. CROPS LIVESTOCK DESTROYED ALONG WITH FACTORY. OVERSEAS ACT OF HUMANITARIAN HELP WOULD PRODUCE DESIRABLE RESULTS ENTIRE POPULATION. MAY I REQUEST YOU MENTION MY INTERVENTION IN CABLE REPLY. KNIGHT

The cable was handed in at the St. Pierre telegraph office at nine-thirty Tuesday morning. There it became the center of one of the few pieces of by-play to enliven the day. A white employee of the telegraph company, presumably a Progressive supporter, made copies of the Knight cable. These later turned up at Progressive Party headquarters and in the office of *Les Colonies*. Then, instead of transmitting the cable on one of the two remaining links still open on the island, he sent the message down to Fort-de-France with the request that the Governor be given a look at its contents.

At Fort-de-France, a clerk, presumably a Radical, received the message; he realized that it had not been transmitted and that it was unlikely to be sent once Governor Mouttet had seen it. He promptly sent the message back to St. Pierre with the suffix, "received and cleared." There, the white telegraphist, assuming that Senator Knight's cable had been cleared by the Governor, sent it to St. Vincent, the first link on its four-thousand-mile journey to Paris. Having transmitted it, the operator tapped out to his colleague in Fort-de-France, "inform Governor cable sent as authorized."

The remarkable thing about the cable was not just its contents, perfectly calculated to produce a reaction in even the most indolent civil servant employed in the Ministry, but the astounding political double dealing at the end of its message: "May I request you mention my intervention in cable reply."

The Sympathy of the Government

With his long experience of the Paris political scene, Amédée Knight must have known that these words would rile the autocratic Minister of the Colonies. Even if Albert Decrais had not been the sort of Minister likely to be outraged at the bald attempt to use his name for the political purpose implicit in Senator Knight's request, it would have been unthinkable for him to move outside the protocol of the situation. This demanded that all communications involving a state of emergency such as the Knight cable indicated should be directly through the Governor.

There can be only one reason for Amédée Knight's arrogance. He was so confident of a Radical victory at the polls in five days' time that he had already assumed the mantle of black power. For Amédée Knight, usually a cautious man, it was extraordinary behavior.

It was shortly after midday when the Knight cable arrived at the Ministry of Colonies. It was passed on to Decrais' office in remarkably little time.

Decrais himself was unable to comprehend its meaning, but he sniffed trouble in the last words of the message and in the signature. He knew and distrusted Knight. A glance at the Martinique file and some questions confirmed his fears.

"With an election pending, the situation was delicate," he said later. "To intervene at the behest of a Party Leader would have slighted the Governor's authority and could have been interpreted as action for political gain."

Instead, the Minister swiftly wrote, in a neat copperplate hand, a reply to the cable Louis Mouttet had sent nearly thirty-six hours before. At the same time he recognized the political pitfall of entirely ignoring Amédée Knight's telegram. With a lifetime's experience on the political tightrope, he wrote:

KINDLY KEEP ME INFORMED ERUPTION AND PARTICULARLY LET ME KNOW NAMES VICTIMS HAVING RELATIVES IN FRANCE. WILL CABLE HELP AS SOON AS RESOLUTION CARRIED. STILL ON SUBJECT REQUEST YOU AND SENATOR KNIGHT CONVEY TO POPULATION SYMPATHY

OF GOVERNMENT. NOT HAVING ANY CREDIT FOR AID PURPOSES AT MY DISPOSAL I HAVE IN VIEW PARLIAMENTARY VACATION HAD TO SEEK INTERVENTION INTERIOR AND AGRICULTURE FOR REFUGEES ERUPTION MOUNT PELEE AND I HAVE PARTICULARLY PRESSED MY COLLEAGUES FOR ALLOCATION FUNDS. WILL CABLE WHEN DECISION TAKEN.

As a piece of Ministerial stalling, it was classic. The right phrases were there: the solicitude about relatives, even if it was basically for French-born white; government sympathy, like royal sympathy, was always a safe thing to express. The over-all impression left was that the grand old lady, Metropolitan France, was gathering up her skirts before running to aid her threatened child on the far side of the world. But it would take time.

While it is perfectly true that the French Parliament was on vacation, the Cabinet itself was meeting in Paris. A few telephone calls would have produced all the authorization and money needed to send immediate relief to St. Pierre. Faced with the description of impending disaster in Senator Knight's telegram, the Cabinet could not have refused some aid. But at the time he sent his cable to Louis Mouttet, the Minister had not yet informed any of his Cabinet colleagues of the situation in Martinique.

Decrais had one drawback as Minister of Colonies: a dislike, at times bordering on the pathological, of the black man. Anything which smacked of decolonization produced a violent reaction from him. To him the world was divided into compartments: European quarters, native quarters; white schools, black schools; the rulers and the ruled. Islands like Martinique where black freedom was firmly established were a blight on the name of France. Behind that cable from Amédée Knight, he recognized a further threat. The Senator was apparently playing another of his devious political games; his cable would act as an effective checkmate.

Help had been promised. But no time had been set when that help would be available. In promising aid, the French Government, through its Minister of Colonies, had recognized its responsibility in the matter. And whatever would be said later, nobody could challenge the attitude of proper concern that his telegram expressed. The population, after all, had the sympathy of the Government.

CHAPTER SIXTEEN

A Handful of Francs

IN ST. PIERRE, Father Alte Roche, close to exhaustion but determined to play out his role as volcanic observer extraordinaire, had, after a brief respite, taken up his familiar position in the window of the Presbytery. From there he noted that "since eight o'clock this morning the roaring of the volcano continued almost without interruption. At intervals the air was filled with concussional shocks that told us something terrible was in progress. St. Pierre had been left in almost night darkness. For many days the disturbed conditions of the atmosphere had interfered with the electric illumination, and it was largely by the aid of brilliant flashes of lightning, which came with almost blinding effect, that the terror-stricken inhabitants managed to grope their way through the streets. The ashes and cinders now fell over a wide area, and in it the vegetation had bowed to mother earth under the load of ash and mud that had fallen."

Under the conditions in which he was making his observations, it is astonishing that the Jesuit was still able to appear so detached from the tragedy that was steadily moving to a climax. Nevertheless, his almost bewildering calmness mirrored the attitude of the Catholic Church on the island. This was the twenty-eighth day since Mount Pelée had awakened, and while everybody else—Progressives, Radicals, the Governor—had given their interpretation of the volcano's behavior, the Church had remained officially silent.

A Handful of Francs

Father Roche and the parish priests had done all they could to ease the lot of their parishioners; so too had Curé Mary in sustaining the courage of those remaining in Morne Rouge. But silence had been the only official attitude expressed by the Church's temporary leader on Martinique, Gabriel Parel. That silence was to provoke comment even among priests not normally given to questioning the attitudes of their superiors. Their feelings were crystalized by Father Roche's diary entry: "There is a general feeling that the time has come for the Church to speak, to offer some definite guidance."

The Vicar-General had not been idle; his diary shows that, but a succession of prayers and well-meaning visits to the afflicted areas were not enough. The evidence of his own eyes would have told him that the conclusion he had reached on Sunday, that nobody really knew what was going to happen, could no longer be valid on this Tuesday: it was the day *after* the Guérin disaster and the horror of the shock wave in St. Pierre. He could be forgiven for not wishing to intervene with the Governor after his return from St. Pierre on Sunday evening; forgiveness would be harder to obtain now for his not intervening on behalf of the town in the short time remaining in its life.

From their emergency headquarters in the Sporting Club of St. Pierre, the "Action Committee" issued the instruction that all roofs and walls were to be washed down to remove the ash. The order was broadcast by the Club's servants running through the streets, and was received as evidence of positive action by the town's elders.

Hosepipes, buckets, even saucepans were pressed into service to remove the ash.

The cindery powder immediately turned to sludge which clogged the streets and the sewage system. In a short time St. Pierre took on the appearance of a gray-brown bog. The sludge slipped and slid through the town. Along the water front it was a foot deep, hiding the aftermath of the tidal wave. As the sludge dried in the warm air, it cracked, releasing a sickening stench. Soon the cracks were filled with the steadily falling cinders.

René Cottrell, on his way from the American Residency where he had spent the morning with the Prentisses discussing the possibility of a direct appeal to Mayor Roger Fouché to have the town evacuated, found that conditions were such that he could barely walk. People kept slipping, colliding, even toppling over as they struggled through the sludge.

"Falling into the mud was often revolting, for it had a foul taste, made worse by excrement or possibly something dead; birds, cats, dogs, and even larger animals had all perished in the stuff, and in the heat rapidly decomposed. It was a common enough sight for a man to vomit when he extricated himself from the filth."

The sludge, aided by the steadily falling cinders, had, in his words, the effect "of turning everybody into creatures who had no place in civilization. From head to foot we were all covered in the filth. Many of the faces looked clay-white, like corpses. I passed people who looked as if they had been buried alive and then dug up again."

As he approached the Sporting Club, he saw that the bust of Napoleon by its entrance porch was unrecognizable, buried under a thick layer of sludge. It was Napoleon who had said that "God, besides water, air, earth, and fire, has created a fifth element—mud!"

Inside the Club he asked for Mayor Fouché. He was informed by Robert de Saint-Cyr, President of the "Action Committee," that the Mayor had left the building.

Cottrell went out into the mud again, making his way to the home of his fiancée, Colette de Jaunville. The engagement party had been confirmed. It would take place on the evening of Ascension Day. Then, in the formal atmosphere of the de Jaunville drawing room, the engagement would be formalized and the wedding date announced.

It was to be the island's social event of the year. The Governor and his wife had consented to stay over for the celebrations after attending Ascension Day services. The diplomatic corps would be strongly represented—apart from the British Consul, James Japp, who was still isolated in his Residency. Members of the Ten Families would be there. Fernand Clerc and Senator Amédée Knight had both been invited and had accepted, though with his acceptance Fernand Clerc had written a note: "My wife and I are honored to accept, providing the mountain allows us to."

Picking his way through the mud, René Cottrell could not help but wonder whether the old planter was not endowed with foresight. Conditions were visibly worsening, and if the sludge increased, it would be doubtful if many people would be physically able to get to the de Jaunville estate on Thursday evening.

In his office in the Town Hall, Mayor Fouché drafted an "Extraordinary Proclamation to My Fellow Citizens of St. Pierre." It read:

> The occurrence of the eruption of Mount Pelée has thrown the whole island into consternation. But aided by the exalted intervention of the Governor and of superior authority, the Municipal Administration has provided, in so far as it has been able, for distribution of essential food and supplies. The calmness and wisdom of which you have proved yourselves capable in these recent anguished days allows us to hope that you will not remain deaf to our appeals. In accordance with the Governor, whose devotion is ever in command of circumstances, we believe ourselves able to assure you that, in view of the immense valleys which separate us from the crater, we have no immediate danger to fear. The lava will not reach as far as the town. Any further manifestation will be restricted to those places already affected. Do not, therefore, allow yourselves to fall victims to groundless panic. Please allow us to advise you to return to your normal occupation, setting the necessary example of courage and strength during this time of public calamity.
>
> The Mayor, R. FOUCHE.

He ordered his secretary to have one hundred copies of his proclamation printed on the flatbed press in the basement of the Town Hall and displayed in all public places. Then Roger Fouché went home for lunch.

Tuesday, May 6, 1902

The latest issue of *Les Colonies* was a sellout. Two thirds of it was devoted to Andréus Hurard's eyewitness account of the Guérin disaster at the refinery and to interviews with the few survivors.

Dr. Guérin himself had refused to grant the newspaper an interview after he had been brought in a state of shock to St. Pierre.

To him the newspaper was a muckraking "little political scandal sheet pursuing its election policies under the very threat of the volcano. Three hours after my factory was carried away, at a time when the entire Mouillage district was still in a panic about the tidal wave, they were putting up electoral posters on the walls," he was to say later.

In turn Hurard had described him, understandably, as "depressed and filled with foreboding for the future."

That foreboding was only increased by a sight of Mayor Fouché's "Extraordinary Proclamation." To the old man who had lost his family, his staff, his business, and a large part of his fortune, the words of that proclamation seemed to refer to another world. It was a "crude piece of trickery to get the population to believe there was nothing to fear, to make them think that Pelée after all was not some pagan god seeking vengeance on all whites. It certainly was not that. But equally, there could be no doubt that it was a real danger. It was clear to me that the town was no longer habitable. It was clear, too, that St. Pierre was heading for misfortune."

Dr. Guérin was determined that he would not be a witness to that misfortune. In the middle of the afternoon, using the considerable personal influence he had on the island, he chartered a small fishing boat, the *Medina*, to take him to safety on St. Lucia, twenty miles to the south of Martinique.

As the boat headed out into the roadstead, Dr. Guérin's last view of St. Pierre was of men with arm loads of posters putting up Mayor Fouché's proclamation. Soon they were lost behind the curtain of falling ash.

To Thomas Prentiss, the American Consul, the proclamation was yet further evidence of the madness which seemed to have gripped all those with authority in Martinique. That afternoon in his office, he sat down and committed to paper his analysis of that madness.

"By any standard the proclamation was an astounding thing. For a start, if the population had been filled with 'calmness' and 'wisdom' there would seem little point in issuing the statement. Again, it was a contradiction in terms, referring, correctly, to 'anguished days,' yet dismissing panic as 'groundless.' The assurances offered as to Pelée's future behavior have no bearing on the facts; but that, like all else in the proclamation, is there for political reasons. The references to the Governor are there for the same reason. So far he has done nothing. Nor do I think he will do much until after the elections are over this Sunday. An order to evacuate the town would be exploited by the Radicals as evidence of Metropolitan France interfering with the island's future. To abandon the elections would be unthinkable. The situation is a nightmare where nobody seems able or willing to face the truth."

He sealed the note in an envelope. He took the unprecedented step of addressing it to Theodore Roosevelt, President of the United States of America.

Thomas Prentiss was wrong about one thing. There were people ready to face the reality of the situation. After lunch the staff of the Lycée gathered in one of the classrooms under the chairmanship of their headmaster, Emile Ricci. He sat at one end of the room; Brisedeaux, the bursar, and Mehous, the assistant headmaster, flanked him. The remainder of the staff sat behind desks, thirty-five men and women, for the most part middle-aged, filled with apprehension.

Of the meeting Professor Roger Bordier recalled that "not even the news that two of our staff, Landes and Doze, were still away on service with the Governor's Commission was a reassurance. To the contrary, the absence of the two professors was taken by the teaching staff to signify that there was more to events than we had been led to believe by *Les Colonies* and other sources.

"The headmaster made it clear to us that though the Lycée was empty, we still had a responsibility to our pupils. He asked us all to go out into the town, try to locate as many of them as we could, and persuade them and their families to leave St. Pierre at once."

The request brought dissension into the classroom. Several of the staff argued that by doing this, they would be cutting across the Mayor's request to maintain the status quo.

"It became clear that through either fear or blind stupidity the staff was not going to act as one body. Like most in the town, they looked elsewhere for leadership and decision. One assistant professor said that to leave would be virtually admitting that Pelée had some magical power."

As the meeting broke up without any action being taken on the headmaster's request, Professor Bordier had already recognized the hopelessness of it all. He walked out of the Lycée and made his way towards *Le Trace*, the road that led to Fort-de-France.

Behind him he left friends and colleagues he would never see again. In less than thirty-six hours they would all be dead; the disaster would wipe out the most prominent teachers on the island.

All day there had been a trickle of people following *Le Trace* to safety. Boverat, the Negro house servant of James Japp, the British consul, counted at least a hundred men, women, and children plodding ahead of him through the ash. He avoided them so that they would not see the tears running down his face.

He had been crying since that moment when his master had ordered him to cross the Roxelane River and make his way south. For Boverat the order had been a painful one to obey. He had been "in service" to the diplomat for eleven years. "To leave, knowing that he might be facing a terrible death, was something I never wished to do. I begged to take him with me. But he could not swim, and he feared that we would both drown trying to cross the Roxelane. In the end he ordered me to go, explaining that there was no need for me to make an empty sacrifice. I could not refuse his order. As I prepared to cross the river, he gave me some papers. They were an account of the last few days in the Residency. He said I was not to open them

A Handful of Francs

until after the drama had resolved itself. From the way he spoke, it was clear that he believed there could only be one outcome—disaster."

Stripped naked, his clothes bundled in an oilskin, Boverat swam the Roxelane. From the far bank he looked back at the British Residency.

The Consul was standing by the tall flagpole in the garden. As Boverat watched, James Japp broke out the Union Jack and hauled it steadily up the pole. He turned and saw his servant. For a moment they exchanged looks. Then both saluted the flag.

Now, late in the afternoon, a couple of miles down *Le Trace,* Boverat saw a column of mounted soldiers coming down the road at top speed. They swept past to a junction where *Le Trace* dropped down into St. Pierre. There the soldiers dismounted, blocking the road. Puzzled, Boverat wondered what they were up to. He had not long to find out. From out of St. Pierre he saw a party of refugees approach the troops. They talked for a minute, then turned back toward the town.

The only escape route by land out of St. Pierre had been blocked.

From her balcony overlooking *Le Trace,* Véronique Clerc had watched the soldiers with growing apprehension. Several times refugees had tried to pass them, and each time they were turned back. She wondered how her husband would fare when his turn came to leave the town and return to their hillside home.

Professor Roger Bordier had found a way to outwit the soldiers. He had cut across a stretch of rough ground behind the town, which had finally brought him out on a cart track. Behind him the track had been cut by the mud flow which had destroyed the Guérin refinery. It joined *Le Trace* a hundred yards south of where the soldiers stood guard. As casually as he could, the professor strolled down the track and on to the main road. The soldiers looked indifferent.

He had gone only a little way when a couple of carriages galloped down toward him. They passed by in a dust cloud. But through the swirling cinders Professor Bordier had glimpsed Louis Mouttet in the first carriage.

Twenty-six hours after the Guérin estate disaster, the Governor and his Commission stood at the edge of the mud, hardening under fresh ash falls, and surveyed the scene. Nothing moved over, or on, the morass; in the distance the chimney stack of the buried factory pointed accusingly towards Pelée. The volcano was hidden behind dense clouds of smoke. From time to time there were muffled reports from inside the pall, and flashes of light, but the ash fall had stopped, leaving the air oppressively warm.

The five members of the Commission of Inquiry were huddled a little apart from Mouttet and his principal secretary, Edouard L'Heurre. For the past week L'Heurre had been on holiday in the south of the island, and had hurried back as news reached him of the Guérin disaster. He had arrived at the Residency at lunchtime. He was thirty years old, ambitious, and sharply observant. As Principal Secretary to the Colonial Administration, he virtually organized the daily administration of the island for the French Government. He had managed to maintain the trust of Radicals and Progressives and the respect of his Governor. In the days to come he would play a vital role.

He was shocked at the changes which had come over Louis Mouttet, and later he was to summarize them: "He was clearly showing signs of *malaise*. He expressed no surprise at my return, or interest for that matter. When I told him that news of the Guérin disaster had spread to all the island, he said it was all part of Radical policies to spread alarm. Later when he discovered that Senator Knight had sent a cable to the Minister, he became positively agitated. Time and again he said that the Senator's intervention was meant to be a political embarrassment. I had never seen him in such a state.

"Then a telegram came from Fouché, the mayor of St. Pierre, with details of the proclamation he had issued, and he [Mouttet] became calm again, and almost elated, saying this was clear evidence that there was nothing to worry about. He insisted that the people would follow the advice the mayor had given them and stay calm. Then another message came from the mayor in which he asked for greater authority to deal with the situation. This made him [Mouttet] become excited again, and he said it was clear from this telegram that the mayor suspected that the Radicals were going to make trouble. On an impulse he said he would send troops to the town to reinforce the garrison there and put down any trouble from the outset. I argued that this would lead to even more trouble; the Colonial Administration would be accused of meddling in political affairs, as they would undoubtedly be if the troops interfered with the Radical electioneering.

"At this he became depressed again, and slumped in his chair, looking excessively tired and withdrawn into himself. In the end he agreed that it would be unwise to send soldiers into the town, but they could be posted on the road outside with orders to stop refugees from leaving who might spread needless alarm over the whole island. I explained to him that from my own knowledge the whole island seemed to be already aware of the damage Pelée had caused, and that it was pointless to post troops for that reason, but he was adamant and insisted that I obey his order without question."

The election, the depression, the suggestion that the Radicals were plotting against him, coupled with the indecisiveness, the withdrawing into himself—all these are significant indications that the Governor was moving further and further away from the reality of the situation. The tragedy was that while L'Heurre recognized there was something wrong, he was in no position to diagnose how seriously disturbed the Governor was. His tale of events this afternoon is a painful description of Mouttet, a man sliding down the scale of sanity.

"Having reluctantly sent off the soldiers, with my own instructions to show sympathy, I was ordered by the Governor to summon the members of the Commission of Inquiry. When they arrived, he told them that we would all be going to make an inspection of the area where the mud had buried the Guérin refinery. He made it quite clear to us that terrible though the Guérin disaster had been, there was nothing to suggest that the broader aspect had changed. I was somewhat surprised to find that Professor Landes also believed this. It was clear that the other members would accept what he said. When we arrived at the edge of the mud, M. Mouttet complained that there was in fact little to see, and he almost suggested that it had been a waste of time to come. Even Professor Landes reacted at this, pointing out that nearly 160 people had died; but when the Governor questioned him, he agreed that the mud could have released pressures building up inside the crater, and so reduced the threat of further eruptions."

The scientific basis for Landes' renewed optimism is dubious, to say the least. The volcano was still alive; *anything* could happen. The attitude of Professor Landes is inexplicable for a man of his standing. Far from giving reassurance, the mud should surely have been a warning to him: if mud could follow a new course down Pelée, so could lava. And a lava flow on top of the mud would be a serious threat to St. Pierre. Why did Professor Landes keep silent? Was he so close to the situation that his critical faculties had deserted him? Did he just make an honest mistake? Or was it fatigue and a feeling of helplessness, since no matter what he said, nothing would change? All these have been advanced as possible explanations. But there is another more likely one: a man who can insist on devoting as much space as he did to the damage to the Botanical Gardens in the Commission's Report can hardly have been the best person to assess the situation in human terms.

"The Governor was plainly anxious to return to Fort-de-France, and his mood now had become almost feverish with excitement. He would cable to Paris a full account of the disaster which had overcome the Guérin estate, but explain that it had also produced a blessing by emptying the pressure from the crater. We were about to go when a member of the Commission asked if it would not be wise to go into St. Pierre, pointing out that the cable could be sent from there. The Governor became very angry at this, and said there was no need for him to go to St. Pierre until he was due to be there on the following evening. To go now, he said, would indicate that he regarded the situation as dangerous, when it plainly was not. He urged us to follow him to Fort-de-France, where he would cable his report and a request for money. I asked him how much he would ask for. He said, as if he had not clearly thought of an answer, that he would ask for five thousand francs. I could not believe my ears."

A man attuned to the situation would have realized there was no need to cable for financial assistance. The island's Colonial Administration could draw on fairly substantial sums of money, and in a state of emergency, funds could be requested from the two major banks in Martinique. But back in his Residency that evening, Louis Mouttet wrote a long, rambling cable to his Minister in Paris, giving details of the Guérin disaster, his view that the situation "remains the same," and ending with these words: "IMPERATIVE YOU SEND ME FIVE THOUSAND FRANCS FOR RELIEF."

Five thousand francs was roughly $500. The money would not buy enough food to feed the refugees in St. Pierre one meal.

CHAPTER SEVENTEEN

The Deepening Crisis

THE GOVERNOR'S CABLE was transmitted from the telegraph office in Fort-de-France at 6:23 P.M. Twenty-two minutes later the duty operator received an urgent message from his colleagues in St. Pierre. It read: "CABLE LINK WITH ST. LUCIA BROKEN. BELIEVED SEISMIC ACTIVITY PELEE RESPONSIBLE. ROUTING ALL EXTERNAL TRAFFIC VIA YOU."

Now only one slender cable linked the island with the outside world, the one from Fort-de-France to Dominica.

This evening, traffic was brisk. Among the incoming cables was the one from Albert Decrais, the Minister of Colonies. The outgoing ones included another urgent one for the Minister from Louis Mouttet. It simply asked: "REQUEST YOU PUT SUCHET AT MY IMMEDIATE DISPOSAL."

The telegram caused considerable speculation in the telegraph office. What conceivable reason could the Governor have for commandeering the *Suchet*, a French warship which had just anchored off Fort-de-France?

"Maybe he is going to shell the volcano," suggested one telegraphist.

"Is he mad?" asked another.

On board the *Suchet* Captain Pierre de Bries was stunned by the events of the previous two hours. He had anchored off Fort-de-France at the start of a routine goodwill visit, and hardly had the anchor dropped than his ship

The Deepening Crisis

had been boarded by Edouard L'Heurre. The Principal Secretary brought an order from the Governor that the warship be ready to go "at a moment's notice" to St. Pierre "if matters got worse."

Whether it was the tone of the order or the fact that it had been delivered in so presumptuous a manner, de Bries bridled. Later he recalled: "I demanded to know what matters could warrant such an order. The Principal Secretary replied that the volcano Pelée was in eruption. I asked how a warship could help. He explained that he was as baffled as I was by the request. I said that even if I had wanted to I could not grant it. It would be a matter for the Naval Minister in Paris."

L'Heurre returned to shore only to sail out to the *Suchet* again shortly afterward. This time he brought a written request from the Governor for the ship to be placed under his command, along with a copy of the cable that Mouttet had sent to the Minister of the Colonies.

"I said I would not move until I personally had received authorization from Paris. But to avoid further visits from the Governor's secretary, I said that I had now discovered engine trouble which would take a day or two to mend."

It was a lie which would save the lives of the forty-three members of his crew and himself.

At sea that evening off St. Pierre, some other captains were also making plans that would save the lives of their crews.

On board the *Topaz* Captain Jules Sequin was preparing to sail to Dominica. After the great wave, he had ridden at anchor well out to sea, alongside the *Orsolina* and *Pouyer-Quertier*. Other ships which had arrived after the wave had included the English cable repair steamer, the *Grappler;* a three-masted French schooner, the *Biscaye;* the *Tamaya,* a French bark; *Quora Maria di Pompeii,* an Italian wooden bark; *Diamant,* a French steamer; *Fusee,* a French artillery steam launch; and the American schooner *R. J. Morse.*

The crews of this fleet had watched the flares burning along the water front of St. Pierre, giving warning of *la*

Verette; and when the ash had stopped falling, they had glimpsed what they correctly took to be funeral processions carrying the dead to the mass grave outside the town. Late in the afternoon the fishing boat *Medina*, carrying Dr. Guérin to St. Lucia, had passed through the small fleet, leaving in its wake news of the disaster at the refinery.

"The speculation was endless. A number of captains paid calls on each other, seeking a common course of action. But it was generally decided that at this distance we were all safe from any threat," recalled Captain Sequin later.

He had no reason to stay. His role as messenger boy for the Governor was over. As evening darkened, he gave the order to steam northward. He was parallel to the town when he spotted a tiny fishing boat approaching. Inside were Louis Labutut, stoker, and Joseph du Quesne, overseer, both survivors from the Guérin estate.

The two men told Captain Sequin that they had stolen a boat in a last desperate attempt to get away from a town they were convinced was doomed to die. They reported "near anarchy on the water front" in St. Pierre.

Nobody was at work. A number of shops had been totally destroyed. Further in the town the situation was unreal. Election meetings, mostly of Radicals, were being held while people fought to keep upright in the mud that had filled the streets.

"It was a mad place, and having heard their stories, I was convinced that I had been correct to leave the area," Captain Sequin said later.

On board the *Poyer-Quertier*, Captain Jules Thirion was in conference with the master of the *Grappler*. They studied a chart of the area on the western seaboard of the island. Across it was traced the *Pouyer-Quertier's* search plot for the broken telegraph cables. The arrival of the English ship was welcome news to Captain Thirion. The *Grappler* had a high reputation for locating faults or breaks in the cables which crossed the seas of the world. She had sailed north from St. Vincent during the day, bringing with her the news of the erupting Soufrière.

"The news of the eruption caused further speculation. Some sailors believed that soon every island in the chain would be in eruption, producing the ring-of-fire effect that the Windward and Leeward Islands are famous for. Certainly, it seemed a bad omen that the Soufrière should have erupted so close to the time that Pelée did. At this rate, we would need every cable ship in the Caribbean to mend the telegraph links which would undoubtedly be broken under the strain of all this seismic activity," recorded Captain Thirion in his log.

Now, on the chart that was spread beside the log book, he drew a line from St. Pierre out to sea. He would search the larger area to the south, while the *Grappler* would concentrate on a smaller sea area north of the town.

"It shouldn't be boring for you with Pelée on your port," he had joked, leading his guest into dinner. "From the sea the volcano offers an ever-changing aspect."

For Fernand Clerc, Mount Pelée assumed only one aspect, that of an increasing threat to St. Pierre. That he could still think of the wider aspect of such a threat is some measure of the qualities he possessed. The planter had tried all means possible to enlist support for his belief that the only sensible step was still to evacuate the town.

"I made it clear to all I approached that evacuation was feasible. It would be simple to move the people by road and sea to Fort-de-France. They could camp outside the town, or even further south, where there would be no danger at all. Enough had happened in St. Pierre for anybody to realize that such a move would be a sensible precaution."

Clerc had approached Father Alte Roche for support. But the Jesuit had "gently explained to me that while he privately welcomed such a move, official Church support could be forthcoming only if M. Parel would endorse it." While Father Roche would confide to his diary that the time had come for the Church to speak out, it was clear to Fernand Clerc that the Jesuit was not willing to be in the vanguard of such outspokenness. Nor, it would seem, was the Vicar-General. Fernand Clerc had sent a message to Gabriel Parel urging him to intercede "at this late hour" with the Colonial Administration. He had not yet received a reply.

He had called upon Andréus Hurard, the editor of *Les Colonies*, "to find that he was visibly shaken by the Guérin disaster." Yet when Fernand Clerc had asked the newspaperman to use his editorial power for a campaign to have the town evacuated, Hurard had bluntly refused, arguing that evacuation would only increase the panic and would certainly cost the Progressive Party the Election.

"I could not understand his attitude," Clerc recalled. "He had seen for himself the damage Pelée was capable of, yet he could still only think of the election! I told him that the election was now of secondary importance, and that in many ways it would be a good thing if the poll was suspended until the volcano had settled down. Hurard was shocked by my attitude. He said that nothing could be allowed to interfere with the election, and that if I thought like this, I might as well concede defeat to the Radicals now! No matter how I argued, I could not get him to see further than his political nose. He kept on saying that he had a duty to perform, and so had I, and that was to insure a Progressive victory on the final ballot."

Fernand Clerc had been equally unsuccessful when he had made an appeal to Mayor Fouché. The ink on his proclamation had barely dried, and a call for the sort of help that Clerc proposed went directly against that proclamation with its curiously worded appeal for calm.

"The Mayor made it clear that on no account would he support evacuation, and he urged me to forget the idea as well. He asked me to lend my influence to calming down the population, and I retorted that the whole trouble was that everybody seemed too calm!"

As with Hurard, Fernand Clerc's relations with the Mayor, at best politely distant, had taken a turn for the worse. A more tactful man would conceivably have phrased his appeals for support more tactfully. But diplomacy had never been one of Fernand Clerc's strong points. In the political arena, he found that he could only speak to most people as bluntly as he addressed his employees.

After his brush with the Mayor, he had appealed to the garrison commander of St. Pierre to lend his name to a

The Deepening Crisis

plea for the Governor to evacuate the town. But the soldier, like Father Roche, while having private sympathy for the idea, felt he could do nothing. His attitude is a little easier to understand: he was still waiting for an acknowledgment of the account of the Guérin disaster he had telegraphed to the Governor. He was clearly fearful that he had overstepped the mark, especially in view of the report that the Governor and his party had been seen driving back down *Le Trace* earlier in the evening.

"The episode had clearly worried the commander. He kept on pointing out to me that this was the first time he had not been notified of the Governor's visit to the district, and he could not help but fear that he had fallen out of favor. He was more preoccupied with this worry than with any threat generated by Pelée," said Clerc.

He had next called upon Thomas Prentiss, the American Consul. There he found support—but of a totally unexpected nature.

"The Consul suggested that the only thing to do was to cable the President of the United States, urging him to ask the French Government to order the Governor to evacuate St. Pierre. To send a cable halfway around the world to urge action by a man a mere fifteen miles away seemed an extraordinary way of doing things. But then, Mr. Prentiss was in an extraordinary mood this evening. He was convinced there was a real danger, but equally he felt it was his duty to remain on hand as his Government's representative. His wife, in spite of my own appeals, also refused to leave, maintaining she had a duty, like my own wife, to be at her husband's side. The matter of sending a cable was resolved when news came that the cable office at St. Pierre was no longer in communication with the world outside the island."

The cable *could* have been relayed to Washington via the telegraph office in Fort-de-France. But from Fernand Clerc's words, it is clear that he had little hope that Prentiss' plan would work.

Finally, with the sort of dogged determination with which he conducted his business, he had approached a number of local shopkeepers and businessmen to start a "common petition" to the Governor for the town to be evacuated. But again the response was lukewarm: "they all felt that such a move would damage their business. They could not grasp the possibility that if the mountain continused to behave as it had done, they would have no business left to transact."

Early in the evening he finally admitted the defeat of any hope of his leading the citizens of St. Pierre to safety. Depressed by the fatalism and apathy which surrounded him, he locked up his office for the last time and walked home. At the road junction on *Le Trace*, he was briefly stopped by the soldiers; when they recognized him, they stood respectfully aside.

At home he bathed, dressed in clean clothes, and dined with his wife in his customary silence. Later they had gone out on the balcony and looked down on St. Pierre.

From his own observations, he had acted to alert the town to the danger it faced. He had been rejected. He could be forgiven for believing, as he did on this Tuesday evening, that he had only one duty left. That was to plan for the safety of his own wife and children.

The needle of the barometer he had fixed to one of the balcony rails had started to flicker again.

CHAPTER EIGHTEEN

A Fateful Reprieve

IN THE MIDDLE of the evening, the Vicar-General, M. Gabriel Parel, arrived in his coach. His decision to hasten to St. Pierre was not altogether out of altruistic motives. Mixed with the undoubted compassion he felt for the population was what also might be regarded as professional avarice.

When he heard that the Governor and his Commission had gone to inspect the Guérin disaster area, Gabriel Parel had assumed, reasonably enough in the prevailing circumstances, that they would go to St. Pierre as well; to his chaplain, Emile Tanoux, he had confided an attractive possibility: after touring the town, the Governor was bound to issue some practical financial relief, and according to Father Tanoux later, the Vicar-General made it clear that if relief money was granted, he was determined that the Church should have control of a portion.

The Roman Catholic Church on Martinique, in spite of its imposing edifices, was not wealthy by the standards of similar Catholic dioceses. Any relief money that its Vicar-General decided need not yet be spent could make a substantial difference to the Catholic treasury on the island.

Gabriel Parel did not, however, know that the Governor's estimate of the relief needed, five thousand francs, was about equal to the usual Sunday offering from all of the island's Catholic churches.

He had been halfway to St. Pierre when the two carriages carrying the Governor and his Commission had galloped past, heading for Fort-de-France.

Like the other key figures in the events of the past few days, the Vicar-General sometimes believed what he wanted to. Once again he confided to Father Tanoux, seated beside him in the official diocese coach, a theory for which he had no support at all.

"I believe the Governor has already decided on the distribution of relief and has returned to Fort-de-France to implement his decision," he told Emile Tanoux.

Without waiting for a reply he leaned out of the coach and shouted up to the driver.

"Go as far as you can. It may not yet be too late!"

Gabriel Parel's preoccupation with obtaining part of the non-existent relief money was to cloud his thinking about the real issue of where the Roman Catholic Church stood in relation to the crisis that was moving steadily upon the population of St. Pierre. A year later he would admit his mistake—but by then it would be of only academic interest.

At the road junction outside the town, the coach was halted by soldiers. They came to attention when they recognized the passengers. The troop sergeant insisted that two of his men ride into town with the coach as an escort.

To Gabriel Parel "the company was welcome, for St. Pierre was not as I had ever known it."

The electricity supply was still off; there seemed to be a shortage of candles. In the streets torches cast lurid shadows. For the Vicar-General "there was now something almost ungodly about this town where I had spent so many happy moments."

Before proceeding, he asked the sergeant how long the Governor and his party had spent in the town.

"Imagine my surprise when he said they had not gone into the town at all, but had driven straight from the area of the Guérin disaster to Fort-de-France. I questioned him closely, but he could offer no explanation for the haste, or as to why they had not entered the town. He knew nothing about any plans for relief money to be distributed."

Their coach had barely entered the town when it became bogged down in the sludge.

A Fateful Reprieve

What irritated the Vicar-General almost as much as this was the attitude of the people. Nobody made any effort to help free the coach, even after he made a direct appeal for assistance; "a strange apathy seemed to cling to all those in sight," he said.

It is likely that what he encountered was something resembling a kind of mass shock. For five days the population had been battered and brutalized by Pelée. And springing from the base of the volcano had come the wave which in turn had brought the plague and an ever-mounting roll of the dead. Few people managed to sleep in the warm, cinder-filled air. The days and nights had become a crazy quilt of rumbling, clouds of ash, and madness. The last trace of positive leadership had evaporated when Fernand Clerc had departed from St. Pierre; now political slogan criers roamed the streets, although few took any notice of their cries; enough was too much. The total result was traumatic; people moved, not really knowing why, and spoke words that often made no sense.

In silence, Gabriel Parel stepped out of the coach with his chaplain. They started to walk down into the town, their cassocks trailing through the mud. They had gone only a short distance when Father Tanoux slipped. Nobody moved to assist him until the Vicar-General helped him to his feet. Sliding and slipping, the two men made their way through the oppressive darkness to the Cathedral of Saint Pierre.

The Cathedral, like the town itself, had undergone a change since the Vicar-General had preached there two days previously. Sludge had piled up against the outer walls of the precinct two or more feet deep in places; the surrounding streets were further clogged with carts and small wagons.

Inside the precinct hundreds of people had gathered. Like those in the streets, they too were silent. For them the ground around the Cathedral was a sanctuary. Among the crowd, lit by the glare of torches, the Vicar-General saw familiar faces from Le Prêcheur and Ajoupa-Bouillon. These refugees produced a characteristic reaction in him: he told them that if they trusted in the Lord they would survive. As he had on the previous Sunday morning in the

Place Bertin, he led them in prayer. He blessed them, and then made his way into the Cathedral through a side door. Inside he found a sight that years later was still to bring tears to his eyes.

The dust lay heavily in the air. In places it had snuffed out the fat candles and tarnished their holders. It clung to the organ pipes, the high altar, and the holy statues. But, he remembered later, nobody seemed to mind. The "strange lethargy" he had met outside the Cathedral had also penetrated to its very heart; "it was as if the will to live had left these people."

They filled every seat, every inch of space in the aisles, even squatting on the steps leading up to the high altar. They sat or stood, the biggest congregation the Cathedral had ever known, silent save for the quiet clicking of rosaries. Many had closed their eyes in silent prayer.

The Vicar-General was encouraged to see "that the priests had not failed in their duty, but were everywhere, offering comfort and reassurance. When the congregation saw me, a cry went up that soon came from all parts of the Cathedral. For a moment I could not distinguish the words. Then Father Tanoux whispered to me that they were asking that I should lead them in prayer for divine intervention. I was moved close to tears at their faith in the Lord and his Holy Church. I made my way to the high altar, and after leading them in prayer, I blessed them. Then, filled with emotion, I made my way to the Presbytery, assuring all I passed that they could spend the night in the Cathedral."

In the refectory room a number of priests from parishes all over St. Pierre had gathered. Among them was Father Alte Roche. He, like the others, had hurried to the Presbytery when news of the Vicar-General's arrival in the town had spread.

They stood in respectful silence as Gabriel Parel came slowly into the room. For a long time he stood there, looking at each of them, at their dust-covered cassocks, their drawn faces. Then, nodding to himself, he tiredly waved them to be seated as he walked to take his place at the head of the long wooden table running down the center of the room.

To Father Roche his superior "was a distressing sight. He was clearly shocked by what he had seen, and for a while he sat there unable to speak."

Embarrassed by the sight of that distress, the priests spent an uncomfortable few minutes as the Vicar-General struggled to regain his composure.

Then he listened without interruption as priest after priest gave him situation reports. The story they told had a depressing similarity of chaos, of ever-worsening breakdown in the social and economic structures of the town. As each priest finished his report, the Vicar-General nodded his head before indicating that another Father might speak; the gaps between the ending of one monologue and the beginning of a fresh one grew increasingly long.

Finally, recalls Father Roche, the Vicar-General turned and asked him: "How is Pelée likely to behave?"

"It is impossible to be sure. But it would be unwise to disregard it as a threat to the town," replied the Jesuit.

"Why?" asked the Vicar-General.

"Because it already has behaved in a manner that I, or anybody else, could not have believed possible even a few days ago!"

"Is the Governor aware of this? The Administration?"

"If he is, he shows little sign of it. Forgive me, but there are many of us who feel that it is the Church's duty to lead in this matter!" said the Jesuit.

"How?"

"By ordering the evacuation of St. Pierre!"

"My dear Father, the Church has no authority to do such a thing. Nor could it lend its support to such a move."

Once more there was silence in the room.

"My friends, I concede that the situation is grave." The Vicar-General paused to consider his words. "But it is not the role of the Church to supercede the will of the State in these matters. The matter of evacuation is one for the Governor. I can recommend that he consider it. But I cannot order it in the name of the Church. And even if I could, I feel this is not the right time for the Church to intervene. There has been no official action by the lay authorities. To the contrary, the Mayor has issued a proclamation asking for calm. That is the official view. To counteract that view at this time would not be effective. My feeling is that the right time for me to speak in the name of the Church is after the celebration of Ascension Day in the Cathedral. Those who attend will then see for themselves the seriousness of the situation, and will be more receptive to the suggestion that it would be wise to evacuate the town."

These words reflect the fatal indecision of the Vicar-General. Nothing must be done, or seem to be done, that would bring the Church into conflict with the State.

Thousands of men, women, and children would soon die who could have lived if a firm decision had been taken now to put the Church's authority behind an order to evacuate the town.

Mayor Roger Fouché, alone in his office in the Town Hall, was working late into the evening drafting his "battle plan" for the Ascension Day celebrations. There would be an official banquet and ball the following evening in honor of the Governor and his wife. Four hundred guests would dine and dance Wednesday night away. Then, as dawn broke on Thursday, the elite of Martinique would be carriaged home to sleep before making their next appearance at midday Mass in the Cathedral of Saint Pierre.

This year an additional significance had been invested in the event. The celebrations would mark the official welcome to St. Pierre for Louis Mouttet and his wife, seven months after their arrival on the island.

Not even Pelée was to be allowed to disrupt that welcome if Mayor Fouché could help it. He realized that "ash would be the enemy. It was staining everything." Yards of white linen would be used; all the silver and tableware would be covered with it until the last moment. The chairs would also be draped with cloth until the guests actually arrived in the large dining room of the Hôtel de l'Indépendance.

"Extra staff are to be detailed to give the floor a last minute brushing. The windows are to remain closed. Adequate staff must be on hand to keep the air cool by hand-fanning."

Fouché turned his attention to the menu. He initialed approval of the eight courses, ticked acceptance of the list of wines to be served with each course, approved the selection of liqueurs to be offered with coffee.

Next to "music" he penned, "As usual." It meant that the string orchestra, recruited from members of the St. Pierre Town Band, augmented by two violinists from Fort-de-France, would stick to its familiar routine of waltzes and quadrilles.

Finally he spread a large sheet of paper on the desk. This was the master seating plan for the banquet. Each of the four hundred places was blank. With the air of a man passing a social sentence, the Mayor started to fill in the tiny squares; a man's standing in the community would depend on the square in which his name appeared.

The members of the "Action Committee" and their wives were demoted from their usual places at the head table to sit among the town's counselors and lawyers. They would not know that they had been deliberately socially downgraded until they consulted the seating plan on the following night. Then they would make one of two choices: they could stage a walkout, or risk public humiliation. Either way Mayor Fouché would have won another round in his clash with the "Action Committee."

In the drawing room of her family home outside St. Pierre, Colette de Jaunville was in the middle of her first lover's quarrel. Like many other such quarrels, it had escalated from mild disagreement to a point dangerously approaching that of no return.

The first rankling of disharmony had come when her fiancé, René Cottrell, had arrived at the house at lunchtime covered in mud and quite concerned about the row between the Mayor and the "Action Committee." While her parents had listened attentively, Colette had found the whole thing "rather boring"; she couldn't understand how her sweetheart could be thinking of such things when there were so many plans to be completed for the forthcoming engagement party. When she suggested that René should clean himself up and join her for lunch, he had replied that he planned to return to St. Pierre to see how the crisis was developing. Colette had been visibly upset. All afternoon she had brooded, in the way that a spoiled girl will, over René's neglect. He had returned to her in the early evening, caked in sludge. Colette's personal maid had shown him to a bath; then, wearing trousers and shirt belonging to M. de Jaunville, he had joined Colette in the drawing room.

Once more he tried to interest her in what he had seen in St. Pierre until, he recalled later, Colette had burst out: "Why don't you spend all your time there!"

"I have spent enough time there to know that it is no place for either of us now," he had replied.

"We are going there tomorrow night to the mayor's ball."

"It may be too dangerous."

"You will not take me?" asked Colette, who now seemed cold and distant.

"The town is like a battlefield. There is mud everywhere. Until you have seen it, you cannot imagine the situation," he replied.

"Are you or are you not taking me to the Ball?"

"I will take you only if it is safe."

"And who will decide that?" she asked coldly.

"I will!"

They had been seated on a divan. Now, Colette rose to her feet and flounced across to a window. For a time she stood there looking out into the darkness. Away in the distance the sky over Mount Pelée was slowly but perceptibly starting to turn red, almost as if it was keeping pace with her steadily mounting anger. Finally she turned and faced René.

"If you will not take me, I will find another escort!"

Without waiting for an answer to her ultimatum, she turned and ran from the room, tears running down her face.

Marguerite, the old washerwoman at the villa of René's uncle, had more prosaic reasons for not wanting him to go to the banquet and ball. René had brought only one dress suit with him from Fort-de-France, which she was now pressing in the villa's ironing room. She had just finished when a film of ash drifted in through the open window, ruining her handiwork. With the stoicism that came from a lifetime of hard won experience, she drew the sacking curtain over the window and started to damp and press away the stains on the suit.

But she doubted if she would be able to prepare the suit on Thursday after the Mayor's Ball in time for the engagement party if Pelée continued to stain everything with ash.

Soon that same suit would be the deciding factor in whether René Cottrell lived or died.

All evening Suzette Lavenière had watched the sky reddening and the lightning forking, and listened to the incessant roaring with something—she was to admit later—close to hysteria.

She sat on the steps of her villa four miles to the south of St. Pierre. Below her in the darkness squatted the estate workers and their families. They all looked to this twenty-four-year-old to provide reassurance. Because they did, she had had little time to dwell on the tragedy of losing her father; she had devoted her time to bolstering their courage.

At dusk they gathered in the courtyard, and she listened as they sang old Creole songs.

When the roaring started, the singing died away and they huddled closer to the steps for protection. Now several hundred eyes watched Pelée.

A few hundred feet below the summit a glowing, roughly circular patch was appearing. At first it was no more than a faint pink, but as the rocks were heated by some vast inner force, the glow deepened. Suddenly the patch leapt into the sky in a trail of fire. The ejected rocks cometed through the darkness before falling into the valleys around Pelée's feet. Through the hole they left high up on the crater's neck came a jet of white hot dust, steam, and molten lava. Immediately the roaring changed to a sound that to Suzette Lavenière sounded as "if all the ship's whistles in the world were being pulled."

The shrill whistling carried plainly to Fort-de-France fifteen miles away. It brought Louis Mouttet hurrying to the window of his study. Edouard L'Heurre watched him with increasing anxiety; all evening the Governor had shown increasing signs "of *malaise*." His conversation had been a jumble of elation, depression, and suspicion, interspersed with long periods of silence. He had narrowed his perspective down to one particular channel for all his stress: the refusal of Captain Pierre de Bries of the *Suchet* to put the warship at his disposal.

As the whistling sound came across the night air, Louis Mouttet had no doubt as to what it meant. Turning to his Principal Secretary, he shouted: "It is the *Suchet* preparing to sail!"

Later Edouard L'Heurre recalled: "It took me some while to convince M. Mouttet that if he looked towards the harbor he would see the *Suchet* lying silently at anchor. Besides, her captain would never have sailed knowing about the cable sent to Paris."

The Governor turned back into the room demanding to know the cause of the noise.

"Pelée," said Edouard L'Heurre flatly.

"Pelée?" echoed Louis Mouttet. "Who has ever heard of a volcano whistling?"

He walked slowly to the desk and picked up a sheet of paper and handed it to Edouard L'Heurre.

"Paris will have to approve this," said the Principal Secretary.

"They will! They will approve *anything* that will insure that the Progressives win the election," replied Louis Mouttet, taking back the paper. "You will cable my decision in the morning to M. Decrais and inform the prison governor."

"And the prisoner?" asked Edouard L'Heurre. "I understand there is great ill will toward him from the other prisoners in the *bagne*. There is a serious risk that they will kill him if he is moved from his cell."

"Instruct the prison governor that the prisoner is to stay in the condemned cell and that no harm is to come to him at least until the election is over. Then I will have him shipped off the island."

A Fateful Reprieve

The compassion that *Les Colonies* had referred to in its editorial the previous Saturday had been exercised by Louis Mouttet. Auguste Ciparis would be allowed to live in the hope that the reprieve would have a decisive effect on the electorate.

But the Governor had one last trick to play in his attempt to sway the voters and discredit the Radicals who presented Pelée as a symbol to bring them victory. He tossed a copy of the Report of the Commission of Inquiry to his Principal Secretary.

"I have already read it."

"And?"

"It is a remarkable report," replied Edouard L'Heurre politely.

"It is more! It is confirmation that Pelée is no threat!"

"But the wave, the damage already done," protested Edouard L'Heurre.

"It is bad, I agree. But the pressures have been taken off Pelée." He turned and walked again to the window. "Even this whistling could be a further reduction in pressure."

"Equally, it could be a sign of a build-up in pressure!"

"Edouard, you are a pessimist. You look for trouble where there is none. Leave that to the experts. Look at the report. There you have the expert opinion of Professor Landes."

"But this report could even now be out of date!" argued Edouard L'Heurre.

"It is not! And furthermore, I have decided that it can only do good if the population of St. Pierre is fully acquainted with its contents. I have sent a copy by messenger to Hurard with instructions that he is to publish a summary of it in the form of an interview with Professor Landes," Louis Mouttet said.

"Does Landes know of this?"

"No. But you can tell him in the morning. There is no need to fear. All Hurard will do is to transpose Landes' views into an interview."

There can be no doubt that Edouard L'Heurre was

shaken by this ploy. A year later he was to testify that he was "utterly opposed" to the "so-called Landes interview." And yet what could this young, ambitious man have done in the face of an unbalanced Governor who held total power over his career? There can be only one answer: if, on this Tuesday night, the Principal Secretary recognized the potential danger to St. Pierre that Louis Mouttet's attitude suggested, it was his duty to take all steps to avert that danger. He obviously believed that Mouttet was sick, even if he did not know just how serious his mental illness was. Furthermore, Mouttet had shown his cunning and deceit in the way he had used August Ciparis and Professor Landes for the same end result—to insure a victory for the Progressive Party, which would be a passport to a continued easy governorship.

For some time now Léon Compère-Léandre had lain in the back of his cobbler's shop in the Place Bertin, listening to the dull plopping sound hitting against the sturdy door of his bolt-hole. Time had long ceased to mean anything, and he was equally oblivious to the smell of stale air and sweat around him.

But the plopping sound was fascinating. Crawling across the floor, he put his eye to a crack between door and jamb. Through it he could make out a solitary child methodically throwing mud balls against the door.

Léon had been crouching there for some minutes when a sudden, vivid sheet of lightning forked across the square. It seemed to cut right through the child before glancing off the brickwork beside his shop. The child crumpled to the ground, hideously burned and quite dead. A split second later another tongue of lightning leapfrogged across the Place Bertin and out to sea. Behind it came a single roar that shook everything.

Terrified, Léon scurried to the back of his shop and buried himself in a pile of sacking.

The roar had tumbled Auguste Ciparis to the floor by its very suddenness. As he struggled to his feet, the cell was filled with light as a new shaft of lightning stabbed from the sky to the town.

A Fateful Reprieve

He could see that the sky itself over Pelée was now a bright red, casting a lurid glow over several square miles. The roaring was coming from inside the cone itself. Unlike any previous sound that the volcano had emitted, this was constant, like the wind from a giant bellows playing upon red-hot coals.

Against the dreadful noise Auguste Ciparis resumed the monotonous low chanting he had sustained for several hours. He had begun when a party of trusty prisoners had taken away the last of the wreckage of the scaffold in the middle of the morning. For the rest of the day he had stood on tiptoe watching to see whether they would return with new timber to build a platform on which he could be hanged on Thursday morning. When the prison's working day ended and they had not returned, his hope had gained strength. In the early evening three *colon* trusties had come to the cell. They locked the door behind them.

From beneath their shirts they produced the now familiar shortened pick handles and proceeded to systematically beat the Negro, who adopted a tried method of defense: "Experience had taught me that to offer resistance was pointless. It only resulted in greater violence. After the first few blows I sank to the floor, trying to protect my head and groin with my arms and legs, curling myself up into a ball that offered the smallest area. All the same it was small protection, and I could not last out long against the blows. I blacked out; when I came to, the men were locking the cell door. Through the pain I heard one of them say that even if the Governor was going to reprieve me, he [the trusty] would make sure that I would suffer hell in the *bagne* for every day I spent there."

He had let the blood dry on his face and body and resumed his chanting. It was almost impossible to distinguish the words as those of the Twenty-Third Psalm. But now Auguste Ciparis had added a few other words to close the Psalm. They were: "God bless the Guv'nah!"

CHAPTER NINETEEN

࿇

Stars and Stripes

FROM HIS OFFICE Andréus Hurard could hear the familiar, and comforting, sounds of a newspaper being prepared for press. They were a reassuring contrast to the shrill whistling that Pelée continued to emit. Each sound—the rattle of type, the hammering of lead, or the clanking of a flatbed—was an individual one that would only end when the last page had been set and the issue of *Les Colonies* for Wednesday, May 7, 1902, was ready for the press to start turning it out. It would be a small issue, four pages, and in newspaper terms "a special"; its contents would be devoted almost entirely to Pelée. The election, which for nearly a month had occupied the bulk of the paper, would be reduced to a brief mention on the front page and a single column of type on the back given over to a new attack on the Radicals. Even this attack would be linked with Pelée. Senator Knight would be accused, with some justification, of using the volcano as "a cheap vote-catching weapon." But the rest of *Les Colonies* was to contain a series of stories about various aspects of the eruption.

By late evening, three of the four pages had been written, copy-edited, composed, checked in galley form by Andréus Hurard, and returned to the printing room. Now only a small pile of proofs remained on his desk. They were stories for the front page. As he read each galley, he scribbled on a headline and made a layout for the story on the make-up page at his elbow. He made no claim to typographical expertise nor to editorial brilliance in his choice of headings. Smaller stories were slotted into the two right-hand columns of the front page; longer ones were earmarked for the two left columns. By tradition the main news story of the day would occupy the center of the page. In that space Andréus Hurard had already scribbled a headline: *"Une Interview De M. Landes."* The interview would be supported by a feature story headed *"Les Volcans."* Both items would owe their substance to the Report of the Governor's Commission of Inquiry, which had been delivered into the editor's hands some time earlier in the evening.

The latest figures of the dead were summarized in seven lines under the heading "The Dead." The flooded Roxelane was reported under "The Flooding of the Roxelane," and contained this piece of dubious deduction: "The rise was caused solely by the heavy rains on the heights. The water holds, in suspension, all the ashes it amasses on its way, and it is therefore very dark. At the river's mouth great quantities of dead fish were found." There was no mention of the corpses that still drifted down through that mouth. Other paragraphs dismissed the tragedy of Le Prêcheur, a story that was in any case four days old. Another story, tucked away at the bottom of the front page, was headed: "Panic in St. Pierre." The twenty-four-line item began by reviewing the panic and exodus from the town, and ended with these words: "We confess that we cannot understand the panic. Where could one be better off than in St. Pierre? Do those who are invading Fort-de-France imagine that they would be safer there than here in case of earthquake? This is a foolish mistake, and it is necessary to put the people on their guard against it. We hope the opinion expressed by M. Landes in the interview we publish will be convincing to those who are most afraid."

This was an extraordinary thing to write, ignoring Pelée and proposing an earthquake as the real threat. In an earthquake it is quite likely that people would not be any safer in Fort-de-France than in St. Pierre, but with Pelée threatening to erupt, there can be no doubt that Fort-de-France, fifteen miles from the crater, was a far safer place.

All these stories only paved the way for the Landes "interview:"

"M. Landes, the distinguished professor of the Lycée, was kind enough to give us an interview yesterday on the subject of the volcanic eruption of Mount Pelée, and of the phenomena which preceded the catastrophe at the Guérin factory. This is what I gathered from our conversation: At five o'clock in the morning, M. Landes saw torrents of smoke escaping from the upper section of the mountain at the spot known as Terre Fendue. He noticed that the Blanche River was swelling to a volume five times greater than that of its greatest known rising, and that it was carrying down great blocks of rock, some of which must have weighed as much as fifty tons. M. Landes, who was then in the Perrinelle settlement, went to Étang Sec at ten minutes before one o'clock. There he saw a whitish mass descend the mountain slope with the rapidity of an express train and enter the river valley, marking its passage with a thick cloud of white smoke. It was not lava but a mass of mud that submerged the Guérin factory. Later on, it appeared to M. Landes that a new opening existed at the foot of Morne Lenard, and this might be ejecting lava.

"The phenomenon of Monday M. Landes regards as unique in the history of volcanoes. It is true, he says, that muddy lavas form very quickly, but the catastrophe at the Guérin factory was due to an avalanche rather than to a lava flow. The valley below has received the contents of Étang Sec, which broke its dike, dropping mud-thickened waters from a height of two thousand feet. If there was no quaking of the earth under the shock of this enormous fall, it was because the sea acted as a stopper, a plug, or pad.

"According to observations of M. Landes yesterday morning, it would seem that the central orifice of the volcano, situated in the higher fissures, was emitting dusty masses of a black and yellow substance in larger quantities than ever, albeit intermittently. It would be safer to leave the lower valleys and to live at a higher elevation if one wished to be sure of escaping the fate of Pompeii and Herculaneum, and not be submerged by muddy lava. 'But,' adds M. Landes, 'Vesuvius has never had many victims.

Pompeii was evacuated in time, and few bodies have ever been found in the buried cities.' Conclusion: Mount Pelée is no more to be feared by St. Pierre than Vesuvius is feared by Naples."

Andréus Hurard could be well pleased with his work. There would be few who would challenge this detailed explanation or its suggestion that St. Pierre was in no danger from Pelée. He could rely heavily on his attention to detail: "Five o'clock in the morning," "ten minutes before one o'clock," "fifty tons," to give credence to the "interview."

But it is precisely this detail that unmasks the "interview." At five o'clock on Monday morning Pelée was not emitting smoke as Professor Landes went to join the Commission of Inqiury at the Jardin des Plantes. He never went anywhere near the Blanche River until some time after the Guérin disaster, so he couldn't possibly have seen rocks, whether they weighed fifty tons or not, being carried down by the river. The Perrinelle settlement lies at the southern foot of Pelée, and was evacuated some time on Friday afternoon, May 2. Étang Sec is a small lake in Pelée's crater. To send the professor, who by all accounts was anything but an athlete, plunging up the side of Pelée to Étang Sec in conditions which were intolerable was to invest Gaston Landes with the sort of powers given to gods. But even if he had managed to climb nearly four thousand feet in the face of a volcanic eruption, a mud flow, billowing smoke, and hot gas, just to be on hand when the avalanche swept down on the Guérin estate, there is still the question of whether his journey was really necessary. He would have had a better view from the top of Mount Verte with Father Alte Roche.

But, according to the "interview," having got to the top, and climbed down again—a feat which in normal times was reckoned to take a man the best part of a day—the incredible professor was hot-footing it across five miles of rough terrain to stand at the foot of Morne Lenard, south of the crater, where he saw signs of eruption breaking out there (something that nobody else ever reported).

The whole thing, of course, is fantasy. On the Monday Professor Landes spent the day with the Commission of Inquiry, first in St. Pierre and then in Fort-de-France. He would have found it impossible, even if he had been mad enough to try, to make his way to the deserted Perrinelle settlement. Certainly he could never have ascended to Étang Sec from there.

The point of this fiction emerges only in the last few lines of the story. "It would be safer to leave the lower valleys, and to live at a certain elevation, if one wished to be sure of escaping the fate of Pompeii and Herculaneum, and not be submerged by muddy lava. . . . Conclusion: Mount Pelée is no more to be feared by St. Pierre than Vesuvius is feared by Naples."

It was the big lie. But Andréus Hurard had no doubt that it would be believed. He knew that his readers retained a touching faith in the truth of what they read in his newspaper. Placed beside other news items of common knowledge, the "interview" would assume a mantle of truth. For Hurard it was a justifiable gamble. It was unlikely that anybody would actually ask the Professor whether he had made that perilous journey to Étang Sec; even if they did, the editor could assume that the professor would act on the Governor's behalf, and would not deny the story.

By late in the evening, Pelée had stopped its whistling. In the darkness the people of St. Pierre waited for the next sign. Those who could not get a roof over their heads stood or lay in the streets, silent and dazed, not knowing what to do or where to go.

In the American Residency, Thomas Prentiss had won a victory he had never thought possible. He had persuaded his wife Clara to sail on the *Orsolina*. With her she would take the letter to the President of the United States. The current issue of *Les colonies* listed the Italian ship's departure as Thursday morning.

In his diary for this night, Thomas Prentiss reveals that "what finally persuaded Clara was my argument that if the American President was aware of the town's predicament, help would be sent. I persuaded her that it was her duty to insure that my report reached him as speedily as possible."

In his Presbytery, Father Alte Roche struggled to decide whether he should go to the Governor and make a personal appeal, "on all the evidence available," for the town to be evacuated.

He had left the Cathedral of Saint Pierre acutely depressed by the Vicar-General's attitude. Gabriel Parel had made it "resolutely clear" that he would like to keep the Church away from any course that brought it into collision with the State—for as long as possible. After the other priests had left the refectory room, the Jesuit had tried to persuade Gabriel Parel to change his mind, but he had "been adamant" that he would "do nothing until Ascension Day."

As he had done so many times in the past few days, the priest walked across to the window of his room. He had been standing there for some minutes when he realized just what it was that was different about this evening. The night sky held something he had almost forgotten. Here and there the darkness was speckled with stars.

Bewildered by the turn of events, René Cottrell sat stiffly in the de Jaunville drawing room. Before him stood Colette's parents.

"It is impossible at this late hour for you to see her," repeated Mme. de Jaunville. "Besides, she is too upset."

After Colette had rushed from the drawing room earlier in the evening René had waited for her to return. Forty years later he was to recall his feelings as he had sat on the divan: "I was in love, and though I could not understand her attitude, I was willing to assume full blame for the quarrel. I waited a full hour for Colette to return, but at the end of that time, her personal maid appeared and informed me that her mistress was resting."

Discomfited, René had returned to his uncle's villa. There he received sensible advice: "Have some food and a glass of wine, and then return and settle the issue before it gets worse." Without waiting to eat, René hurried back to the de Jaunville estate, to find that the row had escalated to parental level. A servant directed him to the drawing room, where he was left for some time before M. and Mme. de Jaunville came in. M. de Jaunville was a cold, precise man. When René thanked him for the loan of the trousers and shirt, he curtly dismissed it with: "I trust it will not happen again." Mme. de Jaunville was elegant in that slightly affected way French colonial women often are. The moment that the relationship between René and Colette had been officially confirmed by René's asking for her hand in marriage, Mme. de Jaunville had set about planning Colette's trousseau and opening negotiations with René's parents on the terms of the marriage contract. She planned to have the contract signed as soon as possible after the formal announcement of the marriage at the Ascension Day party. She had no intention of allowing a quarrel to come between the definite advantages that the marriage could bring to both families. But while she had considerable elegance and charm, she also possessed a violently erratic temper, "and somehow I must have presented myself in a way that aroused this temper," René was to remember.

Possibly it was his understandable request that he should see Colette that aroused Mme. de Jaunville's ire.

"Besides, the hour is late," added M. de Jaunville.

"Indeed it is, sir. But I am anxious that the matter be settled."

"There is only one issue to settle, and that is whether you will take our daughter to the ball or not," retorted Mme. de Jaunville.

Once more René tried to explain to them, as he had tried to explain to Colette, that he felt the situation in St. Pierre was far too dangerous even to contemplate atending he ball; besides "I argued that if the Mayor had any sense, he would cancel the celebration." But again he met with no success. The de Jaunvilles, living cloistered lives outside St. Pierre, had no real concept of what was going on in the town. By the time rumors and reports reached them, they had been filtered through a retinue of servants and generally bore little semblance to the truth. The de Jaunvilles had long discounted those stories, so it is unlikely they would have paid much attention to any stories now coming out of St. Pierre. Like many other large estate owners outside the town, they lived lives almost exclusively contained within the boundaries of their plantation, except for the weekly visit to the Cathedral of Saint Pierre. When

Colette had attended morning Mass on the previous Sunday, the situation in the town, though grave, had not reached the critical stage which had shocked René. He "recognized that it would be hard for them to understand the situation. What I could not understand was that after I had told them about it, they still seemed incapable of realizing the danger."

Once more M. de Jaunville interrupted René: "The hour is late."

"Indeed it is, sir, and more so if you refer to St. Pierre," replied René.

"Would you also have us cancel the party of Ascension Day?" asked Mme. de Jaunville, her temper rising.

With all the earnestness of a young man aware of danger, but not quite clear how to avoid it, René answered: "That, Madame, is a matter for you and your husband. My own feeling is that it might be best to postpone the celebrations until times are safer."

M. de Jaunville argued that the family had given a party on Ascension Day for generations and it was unthinkable to postpone it. "I felt as if I had committed a social error as he explained how all the plans had been made. The Governor and his wife were coming, so was the American Consul and Mrs. Prentiss. Everybody who mattered had been invited. Even suggesting that they should postpone the party made me appear to them like an outcast. They would no more willingly do that than they would allow their servants to dine at table with them," remembered René Cottrell. "All I could do was to explain that Pelée had proved to be no respecter of plans or conventions."

The atmosphere in the drawing room had become icy during these exchanges; its memory remained with René all his life. For a moment the de Jaunvilles were taken aback by his remarks. Then Mme. de Jaunville replied: "M. Cottrell, I must ask that you leave such matters as the behavior of Pelée to those better qualified to decide. At this time you should be concerned only with the problem of being a good husband for my daughter!"

For René the situation had taken on "the aspect of a farce. All I wished to do was to settle the row between

Colette and myself, but instead I had somehow managed to get her parents against me as well! I decided to concentrate on M. de Jaunville, in the hope that he would see the sense of what I was saying and convince his wife of the same. To him I suggested that it would be wise to leave the district until Pelée had calmed down. I told him that a number of people, like M. Clerc, felt this was a sensible course of action to take."

The suggestion was met with frozen silence.

"Leave? Leave to go where?" demanded Mme. de Jaunville.

"To Fort-de-France. M. Clerc says it is safe there."

"And has he gone there himself? Has he left his plantation, warehouses, and office to hide in Fort-de-France?" asked M. de Jaunville.

"I do not know."

"Precisely!" replied M. de Jaunville.

It was clear, remembers René, that nothing he said would convince them of the impending danger. Pelée was over six miles away, and to them the ash falls were simply "uncomfortable."

"There is no more to say on the matter," said M. de Jaunville finally. "We expect you to escort our daughter to the ball. If you fail to honor your obligation in this matter, I do not see how you can proceed with further plans."

Without waiting for an answer, he led his wife from the drawing room. For the second time that night René Cottrell had been handed an ultimatum.

A servant came and escorted him from the house. Outside, the stars were starting to disappear behind fresh smoke that rose in a column from Pelée before spreading over the sky. It was a few minutes to midnight.

In St. Pierre, Wednesday, May 7, 1902, began quietly. To the population huddled in their homes or lying in the streets, there was nothing to suggest that this would be their last full day to live.

WEDNESDAY

❧

May 7, 1902

CHAPTER TWENTY

Gabriel's Angel

SHORTLY AFTER MIDNIGHT, Yvette de Voissous heard the sound of drumming coming from the mulatto quarter. At first she could distinguish no more than the intriguing monotony of drumheads being pounded with bare hands; then, as the noise steadied in its rhythm, she could hear the stick work of the *racklers* and *shuckers*, and she felt very frightened. On previous occasions the drumming had presaged trouble.

From a window in her rooms in the Rue du Collage, she could see into the mulatto quarter. As she watched, a torch flared, and the drumming rose in sudden ferocity. Later she was to describe the events that followed: "I knew the thing had begun, and that this was no place for Christian people, especially good Catholics. Another torch was ignited, and it seemed to send the drummers wild. Then from the edge of the mulatto quarter, not more than a hundred yards away from where I watched, came the shout of first one *quimboiseur* and then another as the voodoo wizards urged their followers on. By now the quarter seemed to be alight with a hundred torches that converged on the ruins of the wrecked Pont Basin. It was a fearful sight, with all these wild creatures running up and down beside the wreckage. They were making what the voodoo people call 'houses for duppys': lamenting in the cult's manner those of their creed who had died in the last few days.

"Even under ordinary circumstances good Catholics

would have avoided such pagan scenes. But now the *quimboiseurs* had chosen to use the occasion to bring into the open their longstanding battle with Christianity. As they shouted out 'prayers' for the dead, they also uttered accusations that their deaths had been caused by the behavior of the Catholic Church, a wicked lie that could only bring further troubles to a town that had already suffered greatly. After one particularly long burst of shouting, the voodoo people turned and ran up into the town. To reach it, they had to come down the Rue du Collage. They came, and there was something so evil around them that you could also smell it. Just in front of the drummers were the dancers, men and women. As I watched, a woman broke through their ranks and like an animal, just like *that*, started to sing and dance at the same time."

The voodoo chant was like an intoxicating spirit that whipped up the crowd. The drums and the movement of the woman were so close that to the fearful Yvette de Voissous it seemed as if "the drumming was coming from her body." Down the street they came, led by the dancer, who, in the spluttering light from the torches looked like "a fiend from hell." Others were joining her, men and women flinging themselves about; when they fell to the ground, they instantly rose to their feet again, swept along by the ever increasing tempo of the music. Drugged by the noise, many of them threw off their clothes. Women were lifted high in the air, their bodies arched until their heads and heels trailed the ground. As they passed Yvette's lodgings, she saw that many of them were drinking; some were already drunk. In the middle of the throng were three *quimboiseurs;* one carried a trussed goat, the other two held chickens, and the fear Yvette felt increased, "for these were clearly sacrificial animals."

The reflection from the torches now lit up the north side of the Cathedral of Saint Pierre as the crowd came to the end of the Rue du Collage. The only entrance to the Cathedral from the north side was through a stout wooden door. Ordinarily it was used by worshipers who were late for a service and did not wish to brave an entrance through the Cathedral's main doors. Now the voodoo worshipers grouped themselves around it. In their midst stood the *quimboiseurs*, with the animals held high. The crowd fell silent. There was a flash of hunting knives, and blood poured from the animals' throats, which the *quimboiseurs* caught in small tins. A sigh came from the crowd as the priests took a drink from the tins, and then with "a leap and a cry" hurled first the dead animals and then the tins against the Cathedral door. The crowd turned and ran into the night chanting strange words that sent fresh fears through Yvette de Voissous: "Let all corpses which are whole rise up from their graves. We command the undecayed dead to leave the Mouillage Cemetery."

Those words made her fear for the peace of her mother's grave.

Clara and Thomas Prentiss had finished their *cafe brulot* in the parlor when the sound of the drumming brought them to their feet. In the street below, a fantastic procession pranced and danced, singing another voodoo song to the drums. In one of the last diary entries she was to make, Clara Prentiss observed: "As they ran past, they appeared scarcely human, moving like hogs, goats, or dogs. Some even imitated chickens, and a few in the center were a most fearful sight as they moved like demons."

So intent were the Prentisses on the spectacle that they failed to notice that the sky over Pelée was starting to glow again.

After he finished editing, Andréus Hurard had gone for a walk. He enjoyed, he had once said, the peace of the streets in the early hours of a day. This morning, though, the streets presented a different image. Aroused by the drumming, hundreds of refugees were milling everywhere, spreading fresh panic with their rumors. The Place Bertin was thronged with those who claimed to have witnessed the outrage committed outside the northern door of the Cathedral of Saint Pierre; to anybody who would listen they poured out their lurid descriptions of what they had seen (later these would form the basis of reports that went around the world of orgiastic behavior that rivaled Sodom; like many other half truths this particular crop was offered as evidence that St. Pierre deserved its fate).

Although the slaughtered goat and chickens had been removed, most people still shrank from passing the door, stained as it was with still-fresh blood. In the end a priest smeared the blood with a handful of ash.

Two men had slept through the noise. One was Father Alte Roche; the other was Gabriel Parel, asleep in the guest bedroom of the refectory. Three more hours were to pass before their sleep would be disturbed, and then by something far more portentous than voodoo cultists on the rampage.

René Cottrell, confounded and bewildered at the way his love affair with Colette had soured, made his way slowly back to his uncle's villa. He had walked about two-thirds of the way when he noticed two things. Pelée was sending a column of glowing ash high into the air that was being carried northward, away from St. Pierre. Closer at hand, a few hundred yards ahead, a light had appeared in the cemetery where a moment before there had been blackness.

As he watched, another light appeared; with it came the low, insistent thumping of a drum—then another pinprick of light, followed by a fourth, forming a circle that started to revolve round and round as the terrified René Cottrell watched. Other lights joined them, forming a cross, a square, a bigger circle. The drumming never varied its beat; low and insistent, it appeared to sustain the lights by its rhythm.

René Cottrell, a lifetime later, was still to remember the fear he felt: "I was terrified and almost fainting. I knew who they were and why they were in the cemetery. I decided that they would probably have preferred a live victim, and since I was the nearest candidate, I hid behind a bush. I was so close to them that it would have been dangerous to turn and run. With the ceremonies they were conducting, it was likely that they had closed the road behind me. If I tried to escape, the chances were that I wouldn't get far. Voodoo people are famous for the way they can move over the ground; I've heard that they can outpace many an animal over short distances. So I watched and waited for them to leave.

"After a while, the drumming stopped, and there was silence. But I could see they were still there. Their candle lights were darting around all over the tombs. They were probably looking for a suitable victim for their loa, one of their voodoo gods, the one that had brought them to the cemetery. They have a number of gods, who need offerings to placate them. This particular loa wanted a body.

"The candles settled on one tomb where they remained for a while. I heard the sound of stones being moved, then in a short time the candles all straightened up, and they started to move across the cemetery in single file directly toward my hiding place. I don't think I have ever been so scared in my life. The candles came running by me, and I was too scared to have seen very much, but I could see that they were stuck on the voodoo people's heads, hands, and even their feet. In the middle of them a group of men were carrying the coffin they had lifted from the cemetery on their shoulders. It was an awful sight."

The group swept on into the night with their coffin, the symbol of their appetite and belief.

At three-ten A.M., the officer of the watch on the *Pouyer-Quertier* entered in the ship's log that Pelée was "on fire down the northern slope."

In St. Pierre the uneasy population could see that a weakness in the rock strata on the northern side of the crater's neck had brought lava welling to the surface and dribbling down the slope.

Fifty minutes later a single, thunderous explosion came from the crater itself as thousands of tons of molten lava were ejected into the air. Pieces of glowing rock fell only a few hundred yards short of the town's boundaries. Pelée had entered yet another phase.

The explosion jerked Gabriel Parel bolt upright in bed. Clutching at his nightshirt, he had hurried to the window of the refectory's small guest bedroom when another tremendous explosion rocked the air. As he stood there, "stunned by this sudden new fury, I watched the most extraordinary pyrotechnic display. At one moment a fiery crescent gliding over the surface of the crater, at the next, long, perpendicular gashes of flame piercing the column of smoke, and then a fringe of fire encircling the dense clouds rolling above the furnace of the craters. Two glowing craters from which fire issued, as if from blast furnaces, were visible during half an hour, the one a little above the other.

"I distinguished clearly four kinds of noises. First, the claps of thunder, which followed the lightning at intervals of twenty seconds; then the mighty muffled detonations of the volcano, like the roaring of many cannon fired simultaneously; third, the continuous rumbling of the crater, and then last, as though furnishing the bass for this gloomy music, the deep noise of the suddenly swelling waters, of all the torrents which take their source upon the mountain, generated by an overflow such as had never yet been seen. This immense rising of thirty streams at once without one drop of rainwater having fallen on the sea coast gives some idea of the cataracts which must pour down upon the summit from the storm clouds gathered around the crater."

As he watched, a third tremendous explosion came from Pelée, and again thousands of tons of white hot lava were ejected through the storm raging around the summit. This time the trajectory was narrower; with a roar the debis was launched on the already ruined village of Le Prêcheur, and further down, on the mud under which the Guérin factory was buried. Then, like a creeping artillery barrage, it bombarded the northern edge of the mulatto quarter.

From his bedroom window in the Presbytery, Father Alte Roche saw the fires break out in the mulatto quarter. One house, dried to tinder, burst into flames under the impact of the glowing rocks. In minutes half a dozen houses were blazing.

"The terrified cries of the poor people filled the air as they ran from the district," Father Roche was to remember. "It would need a steady mind if a major tragedy was to be avoided."

The houses burnt fiercely on the edge of town, threatening to spread and engulf not only the entire mulatto quarter but the whole of St. Pierre. Then through the panic came a troop of soldiers; the sight of these men running toward the danger acted as a brake on the natural fears of the population. In their hands they carried sticks of dynamite, which they lobbed into the burning houses, blowing out the flames in the explosions that followed.

In the chapel of the Convent of the Order of Notre Dame in Morne Rouge, Curé Mary, the village priest, was leading a strange congregation in prayers for the dying Mother Superior, Sister Anselene. More than fifty of the coalwomen of St. Pierre, as well as the Order's twenty-three nuns, knelt in two rows in the front of the chapel. These tall, lithe women had remained in Morne Rouge ever since they had brought food and vegetables to the nearly starving villagers two days before; they were ready, they had told Curé Mary, to help in any way they could, having abandoned any idea of returning to work on the water front. Matching their words with actions, they had brought further fresh supplies of food to the village, carrying it from the south of the island; in the end Morne Rouge had become a huge larder, with enough supplies to keep it going for weeks. Then the coalwomen, the majority of whom were practicing Catholics, had quietly joined the nuns in prayer in the capel.

Sister Anselene had already passed beyond any known medical help the island could offer. Unconscious, she did not hear the plea for merciful intervention, or the low insistent drumming that soon appeared to be coming from all around Morne Rouge. As it started, the coalwomen quietly moved out of the chapel. If Curé Mary sensed the threat the sound contained, he gave no sign.

At four-forty A.M., the watch officer of the *Pouyer-Quertier* logged: "Pelée's eruption is over for the moment, though the heavens above the crater still glow." A few minutes later he made another note: the Italian ship *Orsolina* had weighed anchor and was moving purposefully in toward St. Pierre, in spite of the flares still burning along the water front warning of the pestilence, *la Verette*.

Aboard the *Orsolina*, Captain Marino Leboffe studied the water front carefully. He planned to anchor half a mile offshore. Then, at first light, he would send the ship's mate, Luigi Contoni, ashore to embark passengers and mail and to obtain sailing clearance from the local authorities. If none were instantly forthcoming, he proposed to sail without them, and "nothing was going to stop me."

Nothing was going to stop the soldiers on duty at the road junction on *Le Trace* from carrying out their duties, either. Anybody who looked like a refugee was firmly turned back into St. Pierre. Several times during the night, singly or in small groups, people had failed to pass the soldiers. Now, as the first signs of dawn lightened the sky, the soldiers saw another group approaching from the mud plain which had buried the Guérin estate: a direction from which they expected no one to come. The group was almost upon them before the troops realized that these were not refugees, but voodoo *bourhousses* on the rampage. They came on like shadows, running so lightly that their feet made no sound. They fell upon the startled soldiers, leaping for their throats with the cords they held in their hands—cords made from the well-cured intestines of previous victims. They were light and had the tensile strength of cello strings. Whipped around a throat, they could instantly choke out life.

Two soldiers died before their companions were able to bring their muskets to bear on the attackers, who by then were scattering, swiftly fading off into the darkness.

All through the night Suzette Lavenière had supplemented the courage of her workers and their families. Now, as dawn came, she staggered to her feet, red-eyed from lack of sleep and the sulphur in the air. As she did so, the sound of drumming came from the bottom of the valley, from the area where her father and seven employees had perished on their horses in the mud flow of nearly a week before.

Cries of fear came from the workers. . .

"They will not harm us," Suzette insisted over and

over again. "Take no notice—and they will leave us alone!"

As she uttered her childlike hope, she saw a string of lights emerging on the far side of the valley and making its way up the opposite *morne*. It took Suzette some time to realize that the voodoo worshipers must somehow have crossed the mud, a feat she had thought impossible.

She watched as the string of lights swiftly scaled the *morne*. At the top it gathered into itself, then was extinguished. Now only the drumming remained. That too faded as the sound dipped behind the hill.

Beyond it was Morne Rouge, and Curé Mary, the island's most outspoken opponent of voodoo.

In Morne Rouge the village's spiritual leader, Curé Mary, was revealing qualities that he probably never suspected he possessed. Shortly after the coalwomen had slipped out of the chapel of the convent, he also had left the praying nuns. Outside he had found the village street deserted. The only sound was that of the drumming, low and threatening.

As he stood there in the sulphur-laden air trying to locate where exactly the drumming was coming from, there appeared at the bottom of the street a gang of men, chanting as they advanced. In their midst were the *rackers* and *shuckers* who, hours before, had begun their drumming in the mulatto quarter of St. Pierre. "I knew that they were probably *voodoo bourhousses* from their chanting and style of drumming. I was not personally afraid for myself, for I am an old man. But I was afraid as to what they might do to any villagers who opposed them, and more important, I was afraid of what they would do to the church and convent. For some days I had heard the reports that the *bourhousses* and the *quimboiseurs* blamed the Holy Church for the eruption, and their appearance in the village could only mean that the church and convent at Morne Rouge had been selected as targets. I knew now, or at least I thought I did, why the coalwomen had left. Good and brave though they were, they no doubt felt they were no match for the *bourhousses*, who, when they saw me, stopped as if to gather into them the forces of evil. Then they came running up the street.

"They had come only a few yards when I witnessed a most remarkable sight. From nowhere the coalwomen appeared. In their hands they carried clubs of all kinds. The *bourhousses* were clearly as astonished as I was at the sight of these women, racing upon them, shouting and wielding their clubs as their ancestors must have done when they did battle generations before. The *bourhousses*, realizing they were outnumbered, turned to flee. But they were also trapped from the rear. A group of hunters had appeared." It was the hunting party that had been roaming the slopes of Pelée after witnessing the mud avalanche's destruction of the Guérin estate. "Trapped on both sides, the *bourhousses* scattered, several of them getting soundly beaten in the process; coalwomen and hunters joined forces to chase the marauders back down in the direction of St. Pierre."

Fernand Clerc had been awake long before dawn on this Wednesday morning. Lying in his bed beside his sleeping wife, he listened to the distant drumming with mounting misgivings. He had gone out to the balcony more than once; all he had been able to see in St. Pierre was the flicker of torches. The drumming, along with the other events of the night, had further convinced him he had been right to turn his back on the town.

As dawn came, this feeling was reinforced by the arrival of his chief clerk Raoul Isambert and his family. He had walked out of St. Pierre, leaving behind a situation that, according to the clerk, was "one of complete turmoil."

Isambert, who lived in the mulatto quarter, had fled his home, taking with him his wife and three children, when the voodoo dancers had drummed their way through the streets.

"He told me," recalled Clerc later, "that though there was panic in the streets, the people refused to leave the town for two reasons. The first was that they would be taken by the voodoo people. The second was that they would come within range of a bombardment from Pelée.

Isambert had no doubt that the greater risk was to remain in St. Pierre."

Their journey from the town was nearly cut short at the outset. At the junction of *Le Trace*, the family just escaped being shot down by the soldiers; the children probably saved them from being fired upon. From the troop sergeant, Raoul Isambert learned of the voodoo attackers. On the pretext that it was essential that his master be instantly informed of this outrage, he had talked the family past the soldiers. On arrival at the Clerc villa, the family was installed with the rest of the domestic staff, and Isambert had joined Clerc on the balcony to give a more detailed report of the situation in St. Pierre.

Later Isambert was to testify that "this dawn opened one of the saddest and most terrorizing of the many days of fear that passed. For as the light came, Pelée resumed its hoarse roaring. Coupled with it were vivid flashes of lightning that went through the clouds. Thunder rolled over the volcano's head, and lurid lights played across its smoking column, as it seemed to prepare itself for a supreme effort. But it was the sight in the roadstead that stunned M. Clerc and me. In the growing light we saw that the shimmering waters of the open sea were loaded with wreckage of all kinds. Islands of debris from field and forest had all been swept down into the roadstead during the night, until now, as far as the eye could reach, I saw nothing but signs of destruction. One could hardly have found a more disheartening opening for a new day."

Through this floating morass a longboat was pulling steadily for the shore from the Italian ship, *Orsolina*.

For Auguste Ciparis the hours of darkness had been the longest he had endured. Spent and broken though his body was, his spirit had been revived by a remarkable occurrence. Sometime during the night he had seen "a vision in which there were two men discussing my fate. One of them was older than the other, and the older man said I was not to die after all." As a piece of clairvoyance this is astounding, since that was the night Louis Mouttet and Edouard L'Heurre had discussed, and settled, the fate of Ciparis.

In his cobbler's shop Léon Compère-Léandre had also managed to ignore the volcano, concentrating instead on another sound, closer at hand. In the stifling darkness of his retreat, it had taken him some time to recognize it as the gnawing of rats.

"I lay there," he was to testify later to the official inquiry into St. Pierre, "and I could feel them crawling over my body. They, like me, were starving. I felt a bite on my arm, and then another on my head, and I realized they were prepared to eat me alive. I had always believed that rats would not attack unless cornered, but these rats *were* attacking."

As dawn arrived, Léon Compère-Léandre, the normally placid animal lover, found himself engaged in a struggle to kill the rats that now ran openly across the floor and up the walls of his shop.

In his office in Radical Party headquarters, Senator Amédee Knight read the latest issue of *Les Colonies* with mixed feelings. As he later recalled: "The fiction the newspaper maintained that there was nothing to fear and the preposterous 'interview' with Professor Landes were clear evidence to me that the Progressives were in a state of panic. Yet, the very tone of the Landes' 'interview' could have a decisive effect on the electorate. The professor was a respected figure, and there were many, among them Radical supporters, who would believe that if he felt there was no need to fear, they should not worry either."

The paper also included various incidental information for its readers, not associated with the volcano or politics. On the back page there was an announcement: "Thursday being the feast of the Ascension, the stenographic courses are postponed until next Thursday, May 15. The adult course which was to have taken place Friday next is likewise postponed till May 15." Further down the page: "Our offices being closed tomorrow, our next number will not appear until Friday."

For Senator Knight the coming days would be no holiday. This Wednesday he would spend in St. Pierre drumming up the electorate. Then in the evening he would travel to Fort-de-France in preparation for Ascension Day, which he planned to spend canvassing support in the south of the island for the Radical Party. He was scheduled to return to St. Pierre early on Friday.

But there would be no Friday for St. Pierre.

Gabriel Parel awoke late after his disturbed night. When he did, he found that a further change had come over St. Pierre. From his window he saw that "the roadstead as far as the eye could reach was covered with floating islets, spoils of the mountain, the forests, and the fields, with trunks of gigantic trees, pumice stone, and wreckage of every sort discharged by the torrents. All those flood waters, black and laden with mud, instead of covering the sea with a muddy coat as on stormy days, in tumbling into the sea, barely tinged it with a light yellow streak, and then disappeared as if they were molten lead."

Sometime between celebrating morning Mass in the Cathedral of Saint Pierre and midday, the Vicar-General came to a decision. He felt there was nothing more he could do in St. Pierre. He "believed it my duty to return home," and "I resisted all persuasion to remain." Though later pressed to elaborate, he steadfastly refused to explain just what "duty" took him away from a town that was in dire need of spiritual leadership. Was he after the relief money he believed would soon be distributed? Or does the answer lie in the last entry of his diary for this day: "Was my good angel guarding me?"

CHAPTER TWENTY-ONE

~

The End of the Affair

AFTER LEAVING THE shore, the longboat from the *Orsolina* was chased by two other boats. Yvette de Voissou seated in the prow of the longboat, was the sole passenge In their anxiety to reach the *Orsolina* before being ove hauled, the longboat crew was paying little attention navigation; already they had collided with several dead ar imals, and once an oar had dipped and dragged to the su face a corpse, "bringing shouts of fear from all on boar and causing the boat to tilt in the most alarming manner. The pursuing crews, anxious to gain ground, were als finding that they had continually to fend off all kinds c flotsam that threatened to capsize their boats.

The chase had begun soon after the longboat ha pulled away from the water front. It was carrying the la sack of mail as well as the last passenger to leave S Pierre.

After the voodoo *quimboiseurs* had committed the desecration at the northern door of the Cathedral of Sair Pierre, Yvette had loaded her baggage on a handcart. A dawn arrived, she had trundled her way down to the wate front. Her intention, she was to admit later, was to "some how get on board the *Orsolina*." Arriving at the shore, sh found it a deserted ruin. The only movement came fror the flares warning of the presence of *la Verette*. As the sk lightened, Yvette's problem was solved with the appear ance of the longboat. "The crew were surprised to see me

They asked me what I was doing, and I said I was trying to embark on the *Orsolina*. One of them, M. Contoni (the ship's mate), said that it must have been divine providence that had brought me to the quayside a day early, as the sailing date had been changed on the Captain's orders, and the *Orsolina* would depart as soon as the longboat had returned. I asked about the other passengers, and M. Contoni said there was no time to wait for other passengers, and that anyway he doubted if they would be ready in time, as nobody on the shore had been told of the new sailing date. Later I was to learn that the Captain had refused to stay a moment longer than he had to in the roadstead in view of the eruption. I was also glad to be gone, for St. Pierre was no place for a Christian. One of the crew stowed my baggage and then helped me aboard, while M. Contoni led the others into the town. They returned a short while later with a sack of mail, followed by two men who were very angry."

The pair were from the shipping agency that handled the *Orsolina*. Their anger, Yvette later recalled, stemmed from the new sailing date: "M. Contoni was adamant that the new sailing orders would be enforced. At this, one of the men attempted to wrest back the sack, and succeeded only in arousing the anger of the crew, one of whom threw the sack into the boat, narrowly missing me."

Even so the affair would have probably fizzled out but for the arrival of an official from the harbormaster's office. By all accounts he was a pompous man, and on hearing of the *Orsolina*'s change of departure, he started to quote a number of regulations forbidding it. He was still going through his rule book of customs and health clearance when the longboat crew began to shove off. The official tried to stop this by holding on to the boat's painter, only to be nearly dragged into the water.

"He started to shout, and in no time at all the dock was filled with other officials shouting for us to return. Seeing that we had no intention of turning back, for which I was thankful, they climbed into other boats and gave chase," recalled Yvette, adding with a certain girlish excitement: "I could not help but feel that they would have to be powerful oarsmen to catch up with the longboat."

In the end the longboat held its lead until it reached the safety of the *Orsolina,* where it was hauled up to the deck. There it was met by Captain Leboffe, who to Yvette was a "powerful figure of a man in a great rage. He gave me the briefest of greetings, and ordered a deckhand to show me to my cabin."

The only surviving record of the scene that followed is in the report to the *Orsolina* owners by Captain Leboffe. According to his report, the longboat was still swinging from its davits when the ship's agents and a variety of officials scrambled on board. The Captain "demanded to know why they had pursued my crew going about their lawful duty as if they were common criminals. The agents protested that I could not change my departure date without permission from Naples. I pointed out that as Captain, I, and only I, had the right to decide when my ship was in danger, and I had come to the conclusion that Pelée was a danger to it. An official from the harbormaster said that I would be in breach of contract if I sailed without loading cargo that I was scheduled to carry. Another official said I could not depart without passengers who had paid for the voyage.

"I ignored all these arguments as to why I could not sail, insisting that my only concern was the safety of my ship, and that every minute I stayed arguing, the risk grew greater. The harbormaster's men said they would summon assistance to stop my sailing. I regarded this as a threat to my ship and to my office as Captain. This I could not allow. I told them that if they remained on board to argue much longer, they would find themselves on the way to Europe. Again the officials ordered me to remain at anchor, and I, in turn, ordered my men to escort them from the deck. At the same time I gave orders for the ship to get under way, and the officials had barely reached their own boats before our anchor was on board."

Shortly after nine o'clock on this Wednesday morning, the *Orsolina* started to edge through the debris of the roadstead toward the open sea. Behind her she left eighteen passengers, including the American Consul's wife, Clara Prentiss.

In his cell at midday, Auguste Ciparis was kneeling and praying. A short while before, the prison governor had come to his cell and told him that after a "careful review," a reprieve had been authorized. His "vision" had materialized. The governor had added that Ciparis was to remain in the cell until he was transferred to France, where he would serve out his sentence.

The decision to keep him incarcerated in the death cell would mean that in just twenty hours time, he would be the only person alive in the prison.

That afternoon Fernand Clerc called his family and household to a meeting in the villa courtyard. He was brief and to the point. He told them that he could no longer guarantee their safety, so that all but essential members of the household were to leave at once for Fort-de-France; his wife and children would remain with him, but would be ready to move off at a moment's notice. A carriage would remain harnessed for this purpose.

By early evening the villa was deserted except for the immediate Clerc family and a handful of servants who steadfastly refused to leave until their master departed. He would spend the coming hours seated before the barometer fixed on his balcony, noting every fluctuation of the needle, and at the same time watching the mighty clouds of dust drifting lazily northward in the still air.

Fernand Clerc, like several others, would wonder why the smoke which appeared to stretch unbroken to the horizon had not been seen as a signal for help. He knew that at any given time there were up to a dozen ships sailing within fifty miles of the island, and the sight of that smoke would surely have provoked some interest from them. Nobody on Martinique knew, possibly because of the severed telegraph connections, that the smoke had been mistaken for that emitted by the Soufrière on St. Vincent.

All day, ships had been steaming to that British possession to carry out rescue operations, completely unaware that the island of Martinique was on the verge of a disaster far greater than anything that would happen on St. Vincent. Yet by a strange quirk of nature, the events of this day on St. Vincent would resemble in many ways the fate of St. Pierre.

The Soufrière on St. Vincent had gone into eruption shortly after dawn. It had sent a column of debris shooting an estimated forty thousand feet into the air. By the time the first of the ships coming to the rescue had anchored off Georgetown, the Soufrière had spewed out two great tongues of black mud which rolled down its side. By early afternoon a considerable part of the island was hidden behind the vast reddish-purple curtain that ballooned out of the volcano, remained suspended for a while over its summit, and then spread across the countryside and the southern seaboard.

The Reverend John Darrell, who kept a careful record of the day's events from his vantage point in the belfry of his church in Kingstown, recorded: "A furious fusilade of stones and boulders, a large part of them intensely heated when they fell, was kept up during most of the time and was doubtless responsible for a considerable loss of life."

The official estimate of the dead was later placed at 1350. Many of them had died because they refused to take shelter from the Soufrière's bombardment.

The Reverend Darrell believed that "not one life need have been lost if all had taken refuge in a basement or cellar, or even behind the shutters of a living room. The tragedy was that the Soufrière, in spite of the terror it evoked in some, was for the most part regarded as a spectacle to be observed in the open. Too often those who observed it died for their curiosity."

It was not curiosity which was to prove so fatal for St. Pierre, but lethargy. There was a hint of it in the behavior in Paris of M. Albert Decrais. On Wednesday afternoon the Minister of Colonies was still trying to find the five thousand francs that Governor Mouttet had cabled for. Just why the French economy should be so hard pressed that it could not immediately find such a trifling sum has never been explained. In the office of the Minister for the Navy, a decision had still to be made whether one of the Republic's warships, the *Suchet*, should be put at the disposal of one of France's more obscure colonies. In the end the office would send no instructions until events had overtaken the situation.

On the *Suchet* itself, a lethargic attitude also seemed to prevail. All day Captain Pierre de Bries had observed the smoke rising from Pelée seventeen miles away from where he lay at anchor in the roadstead off Fort-de-France. Even allowing for the natural pique he felt at the Governor's attempt to take over his ship, the captain's attitude is still somewhat curious. He was apparently aware of the threat a volcanic eruption contained for all in its vicinity, and yet he maintained his station. Though later he was to receive deserved praise for his rescue work, he was never questioned as to why he had not sailed a few miles up the coast to investigate from closer at hand the situation in St. Pierre.

Lethargy prevailed on the ships in the roadstead off St. Pierre. All day, they had remained placidly at anchor, though the departure of the *Orsolina* had caused brief excitement. Then late in the afternoon, the *Pouyer-Quertier* and the *Grapplier* had returned from searching for the broken telegraph cables. The two ships had steamed through the little fleet, the *Grappler* taking up station for the night just north of St. Pierre, ready in the morning to start searching the sea bed near where the mud flow had buried the Guérin estate. The *Pouyer-Quertier* headed south, prepared to resume searching at daybreak off Fort-de-France. This deployment seemed to have the effect of rousing the other ships to action. One by one they left their safe moorings offshore to anchor closer to St. Pierre.

By nightfall, the *Biscaye*, the *Teresa Lo Vico*, the *Maria di Pompeii*, the *Diament*, the *Fusée*, and the *R. J. Morse* were riding at anchor a few hundred yards from the water front of St. Pierre. Later they would be joined by the

Tamaya, a French bark, the *Clementina,* a French schooner, the *Korona,* from Hamburg, and the *Gabrielle,* from Marseilles.

One reason why they chose to anchor in what they must have known was a danger area is that the captains wanted to be close to the shore. It would be easier for them to get to the official banquet and ball in honor of Governor Mouttet and his wife. What they did not know was that there had been a change of plans. The celebration, to the consternation of Mayor Roger Fouché, had been canceled.

The man who was responsible for the cancellation now sat in Fouché's crowded office in the Town Hall listening quietly to the Mayor's fury sweeping around the portrait-lined walls. The more he listened, the more convinced grew Edouard L'Heurre that he had been correct in persuading Louis Mouttet to at least postpone the celebration until after the election. It had taken the Administraion's Principal Secretary all afternoon to argue successfully the case for postponement.

"Reports reaching Fort-de-France during the morning made it clear that it would be a major miscalculation to proceed with the celebrations at such a time. At first M. Mouttet would not accept this. But I argued that politically it would be damaging to proceed with the festivities when the town was in such a state. Again, I reasoned that to proceed would allow the Radicals to claim, with certain justification, that the Progressive Party, who principally would support such an occasion, had gravely misjudged the situation. Instead of concerning themselves with a banquet, the Radicals would argue that their concern should be for the people of St. Pierre as a whole.

"Such an accusation would have a decisive effect on the electorate. During the day I had also discovered that a number of guests had declined to attend, pleading that under the circumstances prevailing in St. Pierre, they would be better engaged in more pressing work. Among those who had declined were M. Parel, the Vicar-General, M. Clerc, Senator Knight, and members of the 'Action Committee,' who appeared to be in dispute with M. Fouché.

The End of the Affair

From information reaching me from St. Pierre, I believed that the true reason why they were not anxious to be abroad at night was because they feared further attacks from the voodoo *bourhousses*.

"On all these counts I strongly urged M. Mouttet to order the festivities to be postponed until a more propitious occasion. At last he agreed, and then I proceeded to persuade him that it was his duty to visit St. Pierre to restore confidence in the population. To this he gave his assent, and said that he not only would go, but would take his wife with him. I conceded that this would be a further reassurance, as the sight of Madame Mouttet could only help to placate the population. She was well known to be a nervous lady, and if she was seen in St. Pierre, then clearly many of the population would gain new courage to face the days to come."

There would be no "days to come." But late in the afternoon, dressed in the uniform of a French Colonial Governor, Louis Mouttet, his wife, and Edouard L'Heurre set off by coach to St. Pierre. If Madame Mouttet was nervous, she did not show it.

On the road to St. Pierre, the Governor had encountered members of Fernand Clerc's household, and he remarked that Clerc must be very sure of victory at the polls if he could send so many assured votes away.

"On arrival in St. Pierre, his mood became somber and morose as the full extent of the devastation became clear. St. Pierre already bore all the signs of a town that had been ruined. Ash was piled everywhere, often a foot deep. It was difficult to steer and certainly impossible to maintain any speed. When people saw our carriage, they just stared, seemingly unaware of who we were. Once or twice M. Mouttet called out a greeting, but he received no reply. A week before, the town had been a bustling place, with all minds centered on the election. But now there was no sign of any political interest. Posters had been covered with ash, and the streets, which until a day or two before had rung with the cries of the slogan carriers, were now silent.

"Madame Mouttet was clearly overcome by the sight of the desolation, while M. Mouttet turned to me and demanded to know why he had not been told before of the extent of the situation. It was of no purpose to explain to him that he had rejected all previous attempts to acquaint him with the facts. The truth of the matter is that nobody who was not in St. Pierre on this day could possibly comprehend how serious the situation was. Though it was still fully occupied, the town was dead. Most of the shops had closed. Houses were shuttered. Those who could not take shelter sat in the doorways, or even in the streets, waiting, for who knows what.

"There were groups of soldiers at certain points, but they, like the civilian population, showed little interest in our arrival. We stopped one group and asked them what they were doing, and they said they were waiting for the *bourhousses* to strike again. It was then that we learned of the attack on the patrol on *Le Trace*. Those soldiers, on their own authority, had withdrawn from the road and were now somewhere in the town. That alone gives some indication as to how discipline had broken down. All this seemed too much for M. Mouttet, who soon withdrew into himself, as he had previously done in Fort-de-France."

It took the coach an hour to reach the Town Hall. Then: "In M. Fouché's office a reception committee awaited us; Professor Landes and the Commission of Inquiry were there. The professor explained that he had come on early to the town to see the latest developments for himself. He was of the opinion that it had deteriorated considerably. He strongly refuted the contents of the 'interview' in *Les Colonies,* and said he would demand a full disclaimer. In view of what we had all just seen, it seemed of small import now. Only M. Fouché appeared not to appreciate the situation. He explained that for the most part he had been in his office working on last-minute plans for the celebration dinner. I was about to tell him that the dinner had been canceled when Professor Landes intervened to say that in his opinion the time had come for the town to be evacuated."

The Governor threw the suggestion open to general discussion. Immediately Mayor Fouché opposed the idea. Evacuation, he insisted, would be a disaster. The Progressive Party would lose the election because their city voters would be scattered, while the Radicals could still count on support from their strength in the hinterland. St. Pierre would never recover as the island's commercial center from such a move. L'Heurre's testimony continued:

"And finally, he said there was the banquet. To evacuate, to even think of evacuating after all this carefull planning, was unthinkable. He appeared not to understand when I informed him that the celebrations were canceled. He turned to M. Mouttet, who nodded his confirmation. At this the Mayor became extremely angry, directing his fury first at Professor Landes and then at me. From his desk he produced what he called his 'plan' to insure that the banquet and ball could proceed no matter what Pelée did. To listen to him was to listen to a man living in another world. Everything, he insisted, had been done to insure that the night would be a success. He had detailed staff to remove any trace of dust from the tablewear and draperies! Others had been detailed to cool the air with fans! The orchestra had been rehearsing all day! It was a catalogue of trivia that bore no relation to the seriousness of the situation in St. Pierre.

"He returned to his 'plan' in the manner of a magician producing a seemingly endless number of surprises. For me the only surprise was that no one had previously suspected that M. Fouché could so readily take leave of his senses. The terrible thing was that he seemed unaware of what was happening only a few yards from where he stood. When I asked him if he had inspected the town, as we had just done, he shouted that he had been too busy preparing his 'plan,' adding that if anything serious had happened he would have been informed. I suspect that the success of the celebrations had possibly become such a personal issue with M. Fouché that he had had little time to contemplate anything else. It was only when I told him that a number of the more important guests had declined to attend that he became suddenly silent. Then he questioned me closely on my information, demanding to know the sources.

"The most constant and conspicious impression I had of him was the marked deterioration of his mind. He had clearly failed to grasp the situation, and his thinking was no longer clear, as if he was in the grip of some dementia. Finally M. Mouttet intervened to say that arrangements should be put in hand at once to inform all the guests of the postponement. In the meantime, as it was too late for Madame Mouttet to return to Fort-de-France, they, along with members of the Commission, would stay at the Hôtel de L'Indépendance in readiness for the Mass at the Cathedral on the following day. On M. Mouttet's orders I was to return to Fort-de-France and await any message from M. Decrais in Paris. I was not pleased at the prospect of a night ride in view of the reports of voodoo activities, but it was better than having to witness the distressing behavior of M. Fouché."

In twelve hours time Edouard L'Heurre was to realize that the Governor's order sending him back to Fort-de-France had saved his life.

During the evening, out in the roadstead, a rowboat had gone from ship to ship carrying the news of the canceled banquet. From the decks of the fourteen ships anchored in a rough crescent before St. Pierre, the town appeared ghostly and deserted. Nothing moved. Even the flares warning of *la Verette* had died away.

Around ten o'clock it started to rain. Within minutes the full strength of a storm swept over the town and seaboard, causing flooding and further havoc.

The storm barely cooled the fury of René Cottrell as he rode his horse down *Le Trace* toward Fort-de-France. His anger had mounted steadily all afternoon, from the time he had been awakened. From the rear of his uncle's villa had come the sound of angry shouting. Marguerite, the aged washerwoman, was standing in the yard screaming abuse in the direction of Pelée. In her hands she held his dress suit, covered with ash and obviously ruined.

"Even if I had been willing to take Colette to the ball, the issue was now settled. I did not have the clothes to attend, and to go in anything but the proper attire would be unthinkable," he was to say later.

Once more he had made his way to the de Jaunville estate to explain the latest misfortune which had befallen him. There he was informed that the family were too busy preparing themselves for the ball to receive any visitors.

To René it was "clear that they intended to disregard the advice I had offered. Being young, I felt I had been made enough of a fool, and I returned to my uncle's villa outside St. Pierre, packed my baggage, taking the dress suit with me, and prevailed upon him to lend me a horse. When I reached home in Fort-De-France some hours later, I retired to my room with a bottle of rum, and in a decent time I was drunk enough to fall asleep."

René Cottrell would not awaken until his father brought him the details of the tragedy of St. Pierre, now less than eight hours away.

He would retain his ruined dress suit as a memento until he died in 1948, still a bachelor.

THURSDAY

∾

May 8, 1902

erator to keep calling St. Pierre, "in the hope that a reply

CHAPTER TWENTY-TWO

"Allez"

FOR HOURS, Léon Compère-Léandre had fought a losing battle with the rats which had invaded his cobbler's shop. Then a new terror had been introduced into the darkness. A surge of water had come up through cracks in the floor. He could only guess at what had happened: the cloudburst had flooded the town's sewer system, already partially clogged with ash and debris. Unable to drain away, the water had been forced back on itself and was entering basements and cellars in various parts of the town. Inside his hideaway, the stench was unbearable. Floating in the sewage were rats. "Mad with fear," he had smashed down the carefully constructed barricade between him and the outside world and stumbled out into a deserted Place Bertin. Behind him scurried the rats.

Weak from hunger, his eyesight affected by six days in darkness, Léon stumbled down toward his home on the Mouillage. The air was heavy with sulphur, and soon it brought tears to his eyes. Several times he turned and looked in the direction of Pelée, the air vent he believed the Devil used for breathing. Now, in the first hours of the new day, the volcano appeared to be still asleep. Not the faintest of glows came from its throat. Near the Hôtel de l'Indépendance, Léon came across a man standing and staring intently at the volcano. It was Professor Landes, who "seemed unaware of my presence, just standing there, looking at the volcano."

When he finally reached his home, Léon found it had

been broken into. All the rooms were occupied by refugees, now all asleep on the floors. The only room which had not been commandeered was a small cellar half filled with lumber. Pulling the cellar's stout wooden door behind him, Léon Compère-Léandre sank down on a pile of sacking, determined in the morning to go to the Town Hall and obtain assistance in evicting the strangers who had occupied his home.

For several hours nothing disturbed the early morning, but at a few minutes after five o'clock the volcano rumbled into life. Within minutes the noise had grown to a tremendous intensity. Then came the smoke, red and heavy.

The sight of the smoke aroused a mixture of excitement and fear in Father Alte Roche—excitement because the smoke indicated to the Jesuit that Pelée was entering a totally new phase; the roaring which accompanied the pall was a "clear sign of the turmoil going on inside the crater." His fear, like the fear of the majority of the population, came from not knowing how this new phenomenon would develop. Anxious to obtain a better view, Father Roche left his Presbytery room for the last time and made his way through the town toward Mount Verte. His is probably the only surviving record of just what the last hours in St. Pierre must have been like:

"By half-past five the sun had risen in its course perhaps twenty degrees above the horizon when the roaring of the dark-shadowed mountain grew and grew. In the streets hundreds of agonized people had gathered to make their devotions in the Cathedral of Saint Pierre and the town's other churches, even though there were some hours to go before the first Mass was said. A fellow priest would take morning Mass at my own church, while I would celebrate midday Communion. I had calculated that there would be plenty of time for me to observe Pelée from the summit of Mount Verte and return in time to conduct the service.

"In the streets the panic had been subdued by the news that the Governor and his Lady were in the town, and even now were asleep in the Hôtel de l'Indépendance. Outside the hotel, people had gathered, possibly waiting for the Governor to emerge. In the meantime they stood and admired his fine carriage and pair. The horses, like all else, were covered in ash, and they stamped their feet nervously. In the Place Bertin a small crowd had gathered outside a recess which had been occupied by a cobbler, who during the night had emerged and disappeared.

"Around the Cathedral hundreds of people pressed, hoping for admittance to a place which was clearly already overcrowded. In the Rue Victor Hugo I passed the offices of *Les Colonies*, closed and shuttered, but in an upstairs window I fancied I glimpsed a sight of M. Hurard. I could not help speculating on what his thoughts were now, for it must have been obvious even to him that, with Pelée more active than it had ever been, it must only be a matter of time before the crisis came.

"My route took me through L'Centre and past the American Residency. Seeing my approach, the Consul and his lady appeared. Both said they planned to attend early Mass in the Cathedral and asked if the place was full. I told them they would be exceedingly lucky to find space, though no doubt the office M. Prentiss held would be of some assistance. We briefly discussed the situation, and M. Prentiss told me that he understood the Governor had decided, after a meeting with the Mayor the previous evening, to evacuate the town, and that he intended to make the announcement after the celebration of High Mass in the Cathedral. We all agreed that the decision had come none too soon.

"M. Prentiss asked me for my opinion as to what would happen, and I said it would be imprudent to offer one, as Pelée was clearly not behaving in any manner known before. The Prentisses looked tired, and they told me that they had spent a sleepless night debating the situation. The Consul was clearly distressed that the *Orsolina* had departed without his wife, and he explained that he had some urgent information to communicate to his Government which could help in bringing assistance to St. Pierre. Though I did not venture to say so, it seemed to me that the fate of the town and of its inhabitants would already be settled long before the American Government could intervene."

Father Roche was correct in this assumption. The volcano was moving toward a climax which was now less than ninety minutes away. What he could not foresee, what nobody could predict, was the totally new and shattering manner in which Pelée was to erupt.

The Jesuit took his leave of the Prentisses. On the outskirts of the town, he saw that the Roxelane River was at a dangerously high new level. Already the force of the flood had breached the wall around the British Residency, and water was swirling into the outwardly deserted building. James Japp, the British Consul, was already a forgotten man. Soon he would die in his study, where he had spent the greater part of the most eventful week in his life, apparently continuing his observation of Pelée to the last.

Suzette Lavenière had managed only three hours sleep in the last forty-eight hours. She had maintained a constant vigil with her workers for the approach of the voodoo *bourhousses*, or for a change in Pelée. Around two o'clock on this Thursday morning, she had finally fallen asleep where she sat, and two women carried her into the house and laid her on a couch. Three hours later Pelée's rumbling had brought her back into the courtyard, where she stood for a while, watching the volcano.

"Its time had come. This I felt sure of. Never had I seen smoke of such color: red, hanging in the air, and on the fringe the smoke was black. Over St. Pierre it was growing dark. Beyond, the sun was still shining. But over the town and the ground rising up to Pelée it was dark. It reminded me of drawings I had seen of Calvary on that Good Friday when our Lord was crucified. I felt sure that the final agony of St. Pierre had begun. I joined the estate workers in prayer, all of us kneeling in the dust, feeling that the darkness held a terrible fate in store for all those enveloped in it. After we had prayed for the Lord to spare us and those in the town, we sat in silence and watched the volcano. All the time Pelée was issuing wreaths of smoke which rose straight up. Then they whirled round in a very sharp line, clearly defined against the very light blue sky. Suddenly the smoke was drawn our way. Someone shouted, 'The volcano is coming!' "

"Allez" 251

It was a cry echoed by a number of people who for the past hour had been taking up observation positions on the *mornes* to the south of the crater. They sat, like watchful birds, waiting for developments. There was almost a holiday air about some of them, dressed in their best clothes and carrying hampers of food. Most of them had set out from Fort-de-France to arrive in time for the Ascension Day service in the Cathedral. Few of them could have had any real idea of the situation in the town, but when they saw the smoke gushing out of Pelée and building up over the town, they abandoned their plans and decided to remain at a safe distance, on the surrounding heights. By six o'clock there must have been a hundred men, women, and children waiting and watching. Soon they would be spectators to a disaster such as the world had never before witnessed.

Among the spectators was a professional observer, Roger Arnoux, a member of the Royal Society of France. He was returning to St. Pierre after a lengthy convalescence in the south of the island. A stout, burly man, he was renowned for his caution. When he saw the smoke building up over St. Pierre, he recognized, as he was to write in the *Bulletin de la Société Astronomique de France*, "a dangerous situation." Later his testimony would be used as the basis for the scientific conclusions about the eruption. He took up a position on Mount Parnasse and waited for events to develop. He did not have long to wait.

The smoke whirling toward the Lavenière estate suddenly twisted back, "like a serpent's tongue after prey," and started to "curl round and round the summit of Pelée. Enormous rocks, clearly distinguishable, were being projected from the crater to a considerable height, so high indeed as to take about a quarter of a minute in their flight. I also recognized fixed fires playing with a brilliant white flame. Shortly after six o'clock, several detonations could be plainly heard which confirmed me in my opinion that there already existed a number of submarine craters from which gases were being projected. These exploded on contact with the air. Even from where I sat, the heat was suffocating, and I was soon completely bathed in perspiration. My nerves became agitated, and I concluded that it was uneasiness at the situation that troubled me. I could only imagine how much worse it must be in town."

To Auguste Ciparis the situation was worse than living in hell. For the past couple of hours he had found himself in a stifling darkness. The pall of smoke hung low over the prison, with the density of fog. It was warm on the skin, and contained billions of particles of grit.

To reduce the volume of dust he was breathing in, Auguste Ciparis took off his shirt, urinated on it, and draped the damp garment around his head. Then he curled up in a corner of his cell, clasped his hands around his knees, and tried to avoid as much as possible the dust floating in through the tiny window grill of his cell. What he could not see was that the dust was also piling up against the door of his cell until soon it had blocked out what little light the grill allowed into it. Soon this would mean the difference between life and death for Auguste Ciparis.

Léon Compère-Léandre had made a vital decision. Awakened by Pelée, he had emerged from the cellar to find the squatters in a state of terror at the volcano's behavior. They had pleaded with him to be allowed to stay, even offering money. He refused their blandishments, insisting that they find shelter elsewhere. It was not, he was to insist later, that he was being "unreasonable"; it was just that "you cannot allow people to take over your home." The squatters refused to leave, but the issue was settled when a woman pushed her way toward the cobbler, holding a caged bird in her hand. "She said that if the bird went outside, it would choke in the air. She said that if she had to go, could she leave the bird with me for safekeeping. I was greatly touched by this, and I said they could all remain until the situation had improved."

With that he returned to his cellar.

To Captain Jules Thirion on the bridge of the *Pouyer-Quertier*, the situation had clearly worsened since his ship had resumed searching at dawn for the break in the telegraph cable. The search had steadily brought the

Pouyer-Quertier back up the coast. By six-thirty the ship was roughly off the southern end of the St. Pierre roadstead. Through his telescope, Captain Thirion could see the *Grappler* searching close to the shore just north of the town. "From time to time she would disappear into the cloud of smoke, and when she emerged she was a little closer to shore—certainly closer in than any of the ships anchored in the roadstead."

As he watched, the great pall seemed to lift over St. Pierre as a sudden wind swept down from Pelée to disperse it. Captain Thirion compared the wind to "the crater emptying its lungs. The wind was warm and reeked of sulphur. When it had passed, the sun shone down on St. Pierre. Pelée fell silent and revealed nothing apart from a glow in the rock about three hundred feet from the summit."

By now the *Grappler* was swinging back on its course in a gentle curve as it began a new search plot. Shortly afterward it fired a green flare and stopped. It was the signal that one of the broken cables had been located a few hundred yards offshore.

To Ellery Scott, chief officer of the *Roraima*, the flare was the first sign of normality since arriving in the roadstead; it was comforting to know that "routine work" was being carried out in a situation he had never experienced before. For him, as well as the forty-seven crew and twenty-one passengers, events had first moved out of their smooth routine at five o'clock that morning. Just north of the island, "suddenly and without warning, the ship was engulfed in thick heavy smoke and falling ash." Disturbed by this fallout—"I had never seen an active eruption before, but years ago I saw Etna aflame, so I knew something about volcanoes"—Scott had awakened Captain George Muggah and "asked what he thought of the weather." What the Captain might have thought, not only of the weather but of being awakened at such an hour, is not recorded; what is known is that Captain Muggah gave instructions to remain on course for St. Pierre. The *Roraima* skirted the island, keeping about two miles offshore, "but on account of the currents that were setting us in toward the land we had to steer various courses, sometimes drawing off and at others drawing in. The current was never steady. It ran terribly strong, and we took it for granted that this was due to some volcanic action going on. To a certain extent the Captain and I were alarmed. It was a fine dust, a sharp gray ash that was falling." At six-fifteen the *Roraima* anchored off St. Pierre.

Though he may not have known a great deal about volcanoes, Ellery Scott certainly knew how to give compelling testimony, as this account of the next hour reveals: "The harbor-master and doctor soon came alongside and passed the ship. The next to come aboard were our company's agents, Messrs. Plessoneau and Testart. The Captain had a talk with them and asked whether they thought there was any danger from the volcano. Dense columns of smoke were then rising majestically from the peak of Mount Pelée and ascending toward heaven. The agents were very reassuring. There had been no damage done since the destruction of the sugar refinery a few days before. But Plessoneau and Testarte both said that a number of people wanted to get away speedily to St. Lucia. Captain Muggah decided that it was wisest for us to stop where we were for the moment and discharge the cargo for Martinique. The reason that we had not got to work on it earlier that morning was because, being Ascension Day, there were special services in all the churches of the city. Grand Mass was being said in the Cathedral, and the rich people had come over to St. Pierre to attend it. Laborers and everybody else were religiously inclined for that day.

"Meantime our sailors, under the boatswain, were cleaning up the sand and dust, which lay fully a quarter of an inch thick over everything—just like white sand. The ship was covered with it from end to end. It had sifted into everything. When the Captain and I came off the bridge, our uniforms were completely covered with it. Passengers and crew were gathering up the sands and ashes to keep as mementos. Some would put it in envelopes, others in tin tobacco boxes, and I remember a big Negro giving me a cigar box filled with it, which I took, little thinking what a plenty I would have of it before I made home again.

Meantime the officers were grouped forward on the deck enjoying the grand view of Pelée. The sun was now shining out nice and bright. Everything appeared to be pleasant and favorable."

It was twenty minutes to eight o'clock.

At that moment Captain Edward Freeman of the *Roddam* was dropping anchor a few cable lengths from the *Biscaye*. He noted that nearly all the ships were dressed fore and aft in honor of the feast day, and he gave instructions that once the berthing formalities had been completed, the *Roddam* should run up her own bunting.

The ship had sailed from St. Lucia, south of Martinique, at midnight. Driving through a raging storm, Captain Freeman had several times wished he had not decided to depart for St. Pierre earlier than planned to find out "for ourselves the truth of rumors we had heard on St. Lucia that Martinique had been isolated from the world." As he approached St. Pierre, he became more uneasy: "I did not like the look of Pelée. I asked Chief Officer Campbell whether he thought we should stay clear of the island. We decided to approach the harbor, reasoning that as other steamers were there, it was an indication it was safe, as they would know the situation better than we would."

At twenty minutes past seven the *Roddam* entered the roadstead and received flagged instructions from the harbormaster's office to anchor at the southern end of the natural harbor. As he guided his ship slowly to its anchorage, Captain Freeman noted "how pretty and gay St. Pierre still managed to look in spite of the ash. The Cathedral glistened in the sun. Then from its towers came the sound of the bells summoning the faithful for morning Mass."

They were destined never to complete their act of worship.

The ringing carried clearly, as it had done on more peaceful mornings, to Fernand Clerc seated in a chair on his balcony, intently watching the barometer. Below in the courtyard a servant held a horse and carriage in check. Inside sat Véronique Clerc and her children. They had been there since five o'clock.

"Suddenly, as I watched, the barometer needle moved as it had never done before, swinging wildly. I looked to Pelée. A huge area of rock was glowing below its neck. I knew the climax had come. I grabbed the barometer, ran from the balcony to the coach, and ordered the driver to make all speed from the area."

His gold fob watch read ten minutes to eight o'clock.

Roger Arnoux had been joined by a number of other people on Mount Parnasse during the past hour. Some of them had been about to descend and make their way into St. Pierre when the wind had gusted down from Pelée's summit. As Roger Arnoux described it, "the trees were bowed to the ground; on the higher slopes the ash-laden leaves were torn from the branches of the trees, and some of the smaller branches even broken." When it had gone, he noticed that the glowing rock strata around the collar of the volcano were spreading. Around the summit itself vapors chased each other. They were violet-gray in color and were swept upward by tornadic air currents. Somebody expressed the opinion that they were "very pretty." But Roger Arnoux thought differently.

"I knew as surely as I stood there that Pelée was on the verge of total eruption. Only I could not foresee, *nobody could have,* just how devastating that eruption would be."

As he waited, he saw a carriage hurtling down the road toward the *morne*. It stopped, and the Clerc family "literally tumbled out." Fernand Clerc, driving his family up the hillside "not unlike a sheep dog urging its flock to safety," finally arrived beside Roger Arnoux to gasp out words that, while they could not have been immediately understood, left no doubt as to their broad sense: "My barometer, look!"

But the barometer had been broken when he had wrenched it from the balcony.

"Allez"

On Mount Verte Father Alte Roche was weeping for a town that was yet to die. From his position, he could see that the glowing rock was "in direct line with St. Pierre." Instinctively he knew that when the pressures inside Pelée reached the critical point, thousands of tons of rocks would be sent hurtling onto the town four miles below.

Fifteen miles away in the telegraph office in Fort-de-France, M. Garnier Larouche, the Director of Telegraph and Telephone Services on the island, was doing two things at once. With his eyes he was watching the *Suchet* getting up steam, and with his fingers he was tapping out a routine test message to one of his employees in the cable office in St. Pierre.

At precisely two minutes after eight o'clock the single word "allez" was sent over the wire from St. Pierre. It came as a request to see if M. Larouche had finished his message. It was the last word the outside world would hear from St. Pierre.

As Suzette Lavenière watched, suddenly the sun vanished. Its light was replaced "by a glowing red ball which grew out of the side of Pelée with a noise that was truly terrifying. The sky overhead became invisible. This was the end of the world!"

Her workers hurled themselves into the nearest cover while she stood, transfixed, watching the ball growing before her very eyes. For a moment it remained tied by a fiery umbilical cord to Pelée. Then it detached itself, and slowly, almost lazily, it started to move.

CHAPTER TWENTY-THREE

Requiem for the Living

ON MOUNT PARNASSE Fernand Clerc watched the great molten cloud gain speed with "a rending, roaring sound; a continuous roar blending with staccato beats like the throbbing, pulsating roar of a Gatling gun. It grew so dark that I could only be sure of the presence of my wife and children by groping for them and touching them with my hands."

A few yards away Roger Arnoux wept "as the monster seemed to near us, with a mighty wind acting as its forerunner that even at a distance of over four miles threatened to knock us down. We were now in complete darkness, except for the fearsome glow from the cloud which was now approaching the northern extremity of St. Pierre."

Véronique Clerc remembers: "We saw a sea of fire cutting through the billowing black smoke and advancing along the ground toward the town. What could we do? We held each other close. We wanted to die together, and we were waiting for death. It was a moment of anguish. Fear, lack of air, I know not what was the cause of the fearful choking in my throat. It was raining stones and mud, lumps of mud as big as coils of rope. St. Pierre was doomed. Our friends were doomed. Our world was doomed."

Father Alte Roche watched as "a column of fire, which I estimated to be at least thirteen hundred feet in height, descended upon the town. It engulfed the statue of Christ and the cemetery. Then with a great roaring, it encircled the mulatto quarter, leaping over the Pont Basin, moving across L'Centre, where only a short time before I had conversed with the American Consul and his lady. Then it doubled back on itself, traveling the way it had come, but this time extending into the roadstead, reaching out for the ships berthed there. Before my eyes the *Grappler* disappeared, and then the whole town vanished under the great wall of fire."

From his Presbytery in Morne Rouge, Curé Mary "beheld the black vapor leap from the side of the mountain. Looking down on it as it rolled on to St. Pierre, it seemed to me as if all Martinique was sliding into the sea. A great tongue of fire seemed to detach itself from the vapor to lick up all the water from the Roxelane River. The British Government's Residency was engulfed, as was every building around. Only the towers of the Cathedral of Saint Pierre remained untouched, and they only for a brief moment, for the fiery mass enveloped them, too, as it spread itself over all of St. Pierre. The mass was being constantly refueled by a huge stream of fire pouring out of the side of the crater to ravage an already devastated town. The cane fields were on fire, as were the plantations around the town. There must be so many victims, hundreds, possibly thousands, and from here there was nothing to be done."

On board the *Pouyer-Quertier* the wireless operator was transmitting nonstop a brief message: "St. Pierre destroyed by Pelée eruption. Send all assistance." The S.O.S. had been sent at three minutes past eight. This was a minute after Captain Thirion had noted that the fireball's "forepart became luminously brilliant as it approached the sea. In an instant everything was ablaze, and flames shot from seemingly all points of the city as if from a single brazier. Light detonations followed one another in rapid succession, marking the passage of the cloud. Only one flash of lightning was observed, and that seemed to traverse the cloud from the ground upward. The further incidents of the cataclysm were unobservable, inasmuch as the land was immediately veiled in an impenetrable cloud of fiery smoke, and the *Pouyer-Quertier,* itself under threat, was obliged to flee for safety. In the roadstead every ship seemed to be ignited, and a number, including the *Grappler,* had disappeared."

In the roadstead off Fort-de-France, the *Pouyer-Quertier's* S.O.S. had been picked up by the *Suchet.* Critical though the situation was in St. Pierre, Captain Pierre de Bries, while ordering steam to be raised, was reacting in an oddly detached manner. On receipt of the message, he sent a ship's officer ashore to inform the Governor's office of the distress signal, and to say that the *Suchet* was prepared to go "at all speed to St. Pierre." Fortunately time was saved by the officer's being intercepted on the dock by Edouard L'Heurre, who had just been told by an excited M. Larouche that a "dreadful catastrophe has occurred in St. Pierre. All the lines are dead, and disaster must have overcome the town."

The Secretary himself was in a state of understandable shock. It had taken him only a few moments to realize that if St. Pierre had been destroyed, probably the legislative rulers of the island had died with it. He was the only man left with the authority to fill the breach.

According to the law he was perfectly correct in assuming office, though some would argue later that it might have been more tactful to establish the fate of Louis Mouttet and the Commission of Inquiry (one of whose members, Gerbault, was the Deputy Governor) before hand. But on the basis of the S.O.S. the *Suchet* had picked up, and the report from the Director of Telephones and Telegraphs, Edouard L'Heurre, in his new role as Acting Governor of Martinique, took the first steps to cope with the disaster. He ordered the *Suchet* and the other ships in the roadstead to proceed immediately to St. Pierre to afford what relief they could. To the Ministry of Colonies in Paris he sent this urgent telegram:

ALL COMMUNICATIONS WITH ST. PIERRE LOST.
IT IS BELIEVED ST. PIERRE ON FIRE FOLLOWING

ERUPTION THIS MORNING. GOVERNOR MOUTTET ON SCENE SINCE YESTERDAY EVENING. I HAVE SENT ALL AVAILABLE STEAMSHIPS TO SAVE POPULATION. L'HEURRE.

There would be nobody to save. Of that Suzette Lavenière was now convinced. She had not moved since the great ball of fire had detached itself from its cord and rolled down on St. Pierre. To her the "upper flank of the mountain, facing south, had opened. From the gap, at least three hundred feet wide and possibly as deep, came a stupendous roar, forcing a second black cloud to roll out in huge whorls. It mushroomed upward, forming an even blacker umbrella of darkness. This cloud roared down toward St. Pierre, tumbling over and over. One moment it would clutch at the ground, the next it would rise perhaps a hundred feet before falling back to the earth again. It seemed to be a living thing, glowing all the time, while from its center burst explosions that sent lightning-like scintillations high into the darkness. In less than a minute it had joined the first fireball which had demolished St. Pierre; they merged to blot out everything, and certainly to kill all they touched. The whole city was in flames."

The roadstead off St. Pierre had been an inferno since 8:03, as thousands of barrels of rum and sugar had exploded into the water. Scores of people fleeing the fireball had dived into the sea, only to be engulfed in the wall of flaming rum which swept across the water. The *Grappler* was the first ship to receive the full force of the "death cloud." She did not even have time to catch fire before she capsized and sank. Not one of her crew was able to scramble clear. But, as Ellery Scott was to testify later: "They were the fortunate ones." At exactly eight o'clock he had been standing on the quarter-deck with the third mate who, remarking on the "peaceful sight of St. Pierre, gray, but bathed in bright sunshine," had gone to fetch his camera.

Scott never saw him again, for at that moment came "darkness blacker than night, and as the awful thing struck the water, it just rolled along, setting fire to the shore and the ships. The *Roraima* rolled and careened far to port, then with a sudden jerk she went to starboard, plunging her ice rail far under water. The masts, smokestack, rigging—all were swept clean off to two feet above the deck, perfectly clean, without a jagged edge, just like a clay pipestem struck with a big stick. We had started to heave the anchor, but it never left the mud. There we were, stuck fast in hell.

"The darkness was something appalling. It enveloped everything, and was broken only by the burning clouds of consuming gas which gave bursts of light out of the darkness. The ship took fire in several places simultaneously, and men, women, and children were dead in a few seconds. The saloon and the after end of the ship blazed up at once. The *Roraima* was lying with a heavy list to starboard, pointing toward the shore. Hot ashes fell thick at first. They were soon followed by a rain of small, hot stones, ranging in size from shot to pigeon's eggs. These would drop in the water with a hissing sound; but where they struck the ship's deck they did little damage, for the decks were protected with a thick coating of ashes from the first outburst.

"After the stones came a rain of hot mud, lava apparently mixed with water, of the consistency of very thin cement. Wherever it fell it formed a coating, clinging like glue, so that those who wore no caps were coated; it made a complete cement mask over their heads. For myself, when I saw the storm coming I snatched a tarpaulin cover off one of the ventilators and jammed it down over my head and neck, looking out through the opening. This saved me much, but even so my beard, face, nostrils, and eyes were so filled with the stuff that every few seconds I had to break it out of my eyes in order to see. This mud was not actually burning, but it steamed, and there was heat enough in it to dry on the head and form a crust so that it fitted like a plaster-cast.

"I remember that Charles Thompson, the assistant purser, a fine-looking, burly black from St. Kitts, who stood beside me, had his head so weighed down with the stuff that he seemed to feel giddy and was almost falling. When he asked me to break the casing off his head, I was afraid it would scalp him when I took it off. I could feel the heat on my own head very plainly through my tarpaulin covering, and his scalp must have been badly scorched.

"Everybody was not on deck at this time. Some of the passengers were dressing, some still in their bunks. In some cases they were poisoned almost instantaneously by the noxious gas. In others, they were drowned by the water which swept in hot through the open portholes of the submerged staterooms on the starboard side. The darkness was lit only by the flames from the after end of the ship, and the lurid glare from the shore, where some big warehouses had caught fire. Great puncheons of rum continued to burst with loud reports, shooting their blazing contents into the air.

"At this time I went to the lower bridge, feeling my way along, in order to find the Captain. There on the bridge I almost stumbled on a crouching figure with a hideous face, burned almost beyond recognition. 'Who are you?' I cried, for I did not know him, crouched there in the darkness. The man looked up, his face terrible to see. 'Mr. Scott,' he said, 'don't you know me?' I said, 'My God, it's the Captain!' He got to his feet as best he could. Then, seeing one of the boats still left which was hanging in a crippled condition, he wanted to know if we couldn't clear her away. 'Well, Captain,' I said, 'the boat is stove in and no use, and she is jammed so that twenty men couldn't budge her, and we have got no one to help us.'

"Just then Benson, the carpenter, came on the bridge; his principal burns seemed to be on his hands. The Captain ordered that boat to be cleared away anyhow. With a knife I cut the forward davit tackle, but she wouldn't move. She was jammed. It was impossible to get her clear, and when he found that it was impossible, the Captain said 'Mr. Scott, jump overboard and save yourself.' 'No Captain,' I said, 'I won't leave the ship.' 'Well,' said he, 'find out how the ship is and what is the condition of our people. Find out how the women and children are.' After looking around and finding the after end of the ship all on fire, and people burned and dying everywhere, and fire breaking out in several places forward, I went back to report to the Captain how things were. When I reached the bridge he was gone. He had either fallen or jumped overboard to relieve his own suffering, which must have been very terrible.

"There were only a few of us really able-bodied. But it was wonderful to see their heroism. Two engineers who had lost all the skin on their hands were still carrying things about to help us, using their upper arms and elbows. The first thing to be done was to get the fires out forward, for the wind was blowing offshore and raking the ship, so that we should not be cremated alive. Fortunately the water was calm. It appeared as though the thick rain of mud had smoothed the water, but it still swirled and rolled past us, owing to the volcanic currents. The pumps were clogged and wouldn't work, but every man still able to walk did his best. Two of them began to lower buckets over the side, and then, forming a fire line, we passed them up forward and dashed the water at the flames. All this time thick darkness continued.

"Then all at once, about half-past eight, it lifted a little, and we could see the steamer *Roddam* steaming toward us as though coming to take us off. We had no means of knowing at the time that she was almost as badly off as we were, for she had steerage way and came up close enough for us to see that the forward part of her was all right. We took it for granted that she had been out of the line of fire. It looked like a rescue, and we thanked God. Part of the crew got the passengers—women and children—on the upper deck forward, hoping that the *Roddam* would come near enough to take them aboard.

"All at once, not more than one hundred feet away, she stopped. We said 'Well, perhaps she doesn't see us.' I ran at once to the wheelhouse and grabbed a handful of signal lights. Two of them I found were blue lights and one the company's special signal. We set them off, and they burned brightly like fireworks as we tried to attract the *Roddam's* attention and to show her that some living people were aboard. But to our horror the ship slowly backed out into the darkness, leaving us absolutely disheartened. When the others spoke about it to me, I said: 'She has only backed out of the line of smoke. She will come back again and take us off!' But after a while the wind veered south, the smoke started to clear, and we could see nothing more of her."

The *Roddam* had in fact become a floating charnel house. Her decks were littered with dead and dying crew and passengers, many of whose flesh had been burnt away to the bone. One of the few who had escaped the full impact of the blast was Captain Edward Freeman. The moment the eruption began, he had bolted from the bridge to the comparative safety of a lumber room below deck. Wrapping himself in some old flour sacks, he had waited for disaster to strike:

"Above the roar of the blast I could hear the awful shrieks of those trapped on deck. They were weird, inhuman sounds, like the crying of sea birds in distress. I knew that up there people were being roasted to ashes, but there was nothing I could do, nor anybody else could do. In a trice the ash had penetrated every corner of the ship, and my own hiding place was like a furnace. You can imagine what it was like if you think of going to a blacksmith's forge, and taking up handfuls of fine red-hot dust to rub over your hands and face. Finally I could bear it no more, and I made my way back to the deck. It was a shambles. Bodies everywhere, some burning, some throwing themselves into the sea to drown rather than face further agony. I did not know if a crew could be mustered to work the ship.

"I caught sight of the first engineer and told him to go and see if the anchor chain held fast. He replied that he could not move as he was so badly burned. I was helpless. I turned and looked toward the shore, just a couple of hundred yards away, for the ship was drifting all over the place as the blast had in fact uprooted the anchor, and we were a hulk. From end to end the town was ablaze, with the fiercest fires at the north end where the mulattos had lived. In the Rue Victor Hugo I could discern people running with flames clinging to them. They looked like effigies which had been set alight. Above the roar of the flames, their shrieks carried clearly, joining those coming from the deck around me, until a solid wall of demented sound stretched from the shore to the *Roddam*. As I watched, hundreds of people ran into the sea, their scorched flesh sizzling as it entered the hot water.

"Then the *Roddam* was caught up in a new current, and we drifted back into the blackness which hung over most of the roadstead. By now I had a boatswain and five sailors at my service. I placed two men at the wheel, and I told the others to throw every burning thing overboard. I ran down to the engine room, and heard over the pipe that the second and third engineers had survived. Using my elbows to steer the wheel, for the flesh had been completely burned to the bone off my hands and wrists, I tried to get the ship under way.

"All the time the dust continued to rain. It was so hot that the soles of my boots were burned through, and I sent a man to bring me a pair of thick snowshoes that I had bought in Hamburg last winter. They proved to be just the thing. The ship struggled like a wounded monster. Groans and curses and cries for help came from all quarters. But there was nothing I could do. Then from one of my crew came a shout that the *Roraima* was a few cables away. Somehow I backed away from her and started to move out into the roadstead."

On Mount Verte Father Alte Roche watched the *Roddam* emerging from the holocaust with disbelief. That anything could have survived one of the biggest explosions the world had known was beyond his comprehension.

"In three minutes St. Pierre had been completely devastated. The shore was a mass of wreckage, and even from this height I could see that there were bodies everywhere. No one could have survived for longer than a few minutes. But for them they must have been terrible minutes. Every street and every building was in ruins. In the Rue Victor Hugo there appeared to be hardly one brick on top of another. The Cathedral had been decimated. In a moment Pelée had made St. Pierre look as if it had been in ruins for a thousand years. Nothing remained. Not the Cathedral, the Churches, the banks, the hospital, the theater, the Sporting Club, the offices of *Les Colonies*, the Residencies of the foreign diplomats. All had been demolished. St. Pierre was a dead town, filled with dead people."

Requiem for the Living

But, incredibly, there were survivors.

In his cell, Auguste Ciparis could not understand why he was not dead. His cell was half filled with rubble from the prison block which had collapsed on top of him. Through the grill came a jet of scalding air. He had freed himself from the falling masonry only to be knocked off his feet by the force of the blast of superheated air howling through the grill. Suddenly it stopped, and a fearful silence took its place. Ciparis found that his whole body was seared as if he had been placed on a red-hot grill, and he was bleeding in a score of places. The blast which had forced its way through the grill had undoubtedly lost some of its impact under the ash piled thickly against the bars. From beyond those bars now came cries for help.

"I put my face near the grill and saw that the darkness had lifted. The prison walls had gone. So had all the buildings beyond it. I could see," he claimed later, "almost as far as the mulatto quarter. Near the ruined Cathedral the bodies lay thickly. When the explosion occurred, the congregation must have rushed out of the Church into the street. The death cloud had stripped them of their hair and clothing and thrown them down naked in all manner of twisted shapes. From some of them came cries of 'I'm on fire, I'm dying.' Then there was silence, more terrible than what had gone before. I thought all in St. Pierre must be crushed by an earthquake and fire. I smelled nothing except my own body burning. Soon I heard nothing except my own unanswered cries for help."

He would spend three more days in the cell before help came. After being rescued by one of the search parties, he would be taken to Morne Rouge, where Curé Mary would nurse him back to health. Later his sentence would be suspended, and he would earn a living as an exhibit in Barnum and Bailey's circus, spending his days in a replica of the cell in which he was now entombed.

Léon Compère-Léandre had emerged from his cellar when the first full fury of the blast had passed. He faced almost indescribable horror. Most of the refugees had been stripped naked by the force; several had their entrails exposed; a child's face had been completely burnt away. The woman who had offered him the caged bird was burning away in a corner of the room. Her pet and its cage had been completely incinerated. Standing there, Léon felt:

"I was on the fringe of Hell. The house, as a house, was no more. Being on the waterfront, it had escaped the full effect of the eruption. But it had been demolished to the ground floor. Even as I watched, incredible things were happening. I saw my waistcoat hanging on a wall spontaneously catch fire. I moved to help the burning woman, but she gave a terrible shriek and ran from the house in a ball of flame. In the front room I found a man still clothed and lying on a bed, dead. He was purple, and as I watched, his body started to inflate as if it were filled with gas. On the floor another body, that of a younger man, had already burst open, and his gut was everywhere.

"In the corner, I saw two corpses, intertwined, as if the man had been trying to shield the girl. Both were grotesque in their nakedness. I saw another woman, fallen on her back, naked, but very little carbonized. One of her hands was on her breasts, digging into the flesh. The other hand seemed to be protecting the face. An attitude of battle against the flame. But the most terrible injuries were on the threshold of the house, as if the bodies had been trying to flee when the blast came. Two in particular I remember. One, fallen forwards, face down, legs spread-eagled. Between them, the other, kneeling, chest thrown forward, head up. The head was scalped, burned, the eyes gone, the lips formless, surrounding something black which turned out to be a tongue made of charcoal. I was sick all over them, and crazed with the sight of it all.

"It was only some time later that I saw that I, too, was burned; my legs, my arms, my chest, all were raw with burns. I ran from the house, which was now aflame. Around me the entire city was aflame. In the Place Bertin I stopped to urinate, and when I did, the ash smoked. Around me there were hundreds of bodies, all dead. The Hôtel de l'Indépendance was a ruin, and in it must have perished the Governor and his wife. Once I turned, and there was nothing to block my view down to the roadstead. There I could see ships on fire, but I could not help but think their crews were better off than I was."

Léon Compère-Léandre eventually reached *Le Trace*, the road to Fort-de-France, where he was picked up by a rescue party and taken to a hospital in Fort-de-France. Later he would return to St. Pierre as a special constable employed as part of the force guarding the ruins against looters.

The crews aboard the surviving ships in the roadstead would have disputed Léon's belief that they were better off than he was. Along with the *Grappler*, the *Tamaya*, *Clementina*, *Diament*, *Fusée*, *Biscaye*, and the *R. J. Morse* had already sunk, taking with them nearly two hundred men. The *Maria di Pompeii*, *Roddam*, *Teresa Lo Vico*, *Korona*, and *Gabrielle* were all smoldering hulks, with the majority of their crews dead or dying. On board the *Roraima* Ellery Scott had taken over command:

"This was about eight-forty-five. How long we could stay afloat was the question. There was no time for deliberation. All of us who could rushed to the life belts, which were distributed through the ship in various places, and put them around every living soul aboard. When a mother had a child in her arms, we would pass the preserver right around both of them together. The next thing was to find out what condition our battered hull was in and to put out the small fires which had ignited again here and there. The worst one was in the port steerage, far forward.

"The women's quarters, as it happened, had been freshly cleaned, and mattresses were neatly piled inside, while the door was kept locked lest the crew should steal the beds. The starboard ports, however, were left open, and the volcanic fires sweeping in had ignited the mattresses. We tried to open the door. But, finding it fast, several of us grasped a big plank, and making a battering ram of it, smashed the door in. Two great piles of mattresses were all afire. It was a bad outlook, for if the fire gained headway there, it would sweep the ship.

"Worse than this, we had a matter of three thousand cases of kerosene oil, great kegs of varnish, and barrels of tar stowed away in the forward hold of the ship, not a dozen feet from where the fire was. Out on the deck, just over from the steerage quarters, were the cattle pens, which were used to store some ten thousand feet of spruce lumber, enough to burn a city, and standing close by were a number of puncheons of temper lime, a highly inflammable substance used in the making of sugar. The temper lime, which makes fire if it comes into contact with water, was already smoldering, and the smoke was hanging thick about it.

"It was a bad fight, this at number two hatch. Two of us lowered buckets over the side and hauled up water, while the others dashed it upon the mattresses. The water would quiet the flames for a moment, and then one of us would dash in, pull out a mattress, and throw it overboard. But the instant a smoldering mattress came into the current of air outside, it would blaze up again, and it needed lively work to get one clear without being badly burned. All this was exhausting work, but there was more to follow.

"We soon saw that the firemen's quarters on the starboard bow were breaking out in flames. We fought them hard and steadily and again dragged out the mattresses one by one, and more than once as we did so, out with the mattress would come the lifeless body of some messmate who had been trapped like a rat. After a while all the smaller fires came under control, and we got a breathing spell so that we could look about us.

"The sight was fearful. All around us were sailors and passengers, men, women, and children, burned and dying, crying aloud for water. Gradually we collected the survivors and laid them on deck forward near number one hatch. All of them cried for water. But many of them could not drink at all. The flaming gases had burned their mouths and throats and even the linings of their stomachs, so terribly that in many cases the passage of the throat was almost entirely closed. When we put the water into their mouths it stayed there and almost choked them, and we had to turn them over to get the water out, and still they would implore us for more.

"Fortunately the darkness was lifting now, and the flaming city supplied us with all the light we needed. We broke open the icehouse door and hauled out blocks of ice and broke them into small pieces. These the sufferers could hold in their mouths when they could no longer drink. Several of them had their tongues completely burned out. All this time the groans and shrieks of those still with tongues were heartbreaking. You read about the rich man in the place of torment looking up and asking for water. Well, that is about as near as I can come to describing it, but everything that happened sticks in my mind like a nightmare.

"I can see now one of the passengers, a man, lying on the fo'c's'le deck, hideously scarred, crying for water. When we gave it to him, he could not drink it. It would not pass down his throat. He was crawling around on deck on his hands and knees calling for water, and at last we were afraid he would fall overboard, so with the assistance of another man, I brought him down to the main deck. As soon as he was there, he at once began to crawl about like a dog seeking water. The man's tongue was literally burned out of his head. His arms were cruelly burned from the shoulders to his finger ends. But the worst burns were internal. The fire did not seem to penetrate the clothing, but wherever the flesh was exposed, it burned mercilessly.

"I saw one little coffee-colored baby fearfully scorched lying in the arms of a white nurse called Clara, who had come from New York with a family named Stokes. The child was in a dying condition, its tongue lolling out of its mouth and the skin of the tongue all gone. There was still life in the little thing, and as one of the crew gave it some water, it died in his arms. Then he laid it in one of the deck staterooms. The door was open, so that we could all see it, and the sight was so pitiful that I went in, and, shaking a pillow out of its case, put the little disfigured body inside and then laid it on the bed so that it looked decent and Christian. I am thankful to say that Clara survived and later went to a hospital. But before that she had helped us take care of Mrs. Stokes and her three children, two boys and a girl. The wretched woman's mouth could not open, and her teeth were set. We took a small spoon and put some crushed ice between her teeth, and could hear her murmured thanks.

"Poor creature, she did not live long enough to see two of her children die. The elder boy died ten to fifteen minutes afterward. Later we got the baby, who was little more than an infant in arms, and the girl on board the rescue boat. But the baby died before it reached land.

"The women behaved very well, though they were all terribly burned. One big colored woman, for all her burns and scalds, kept singing hymns. Between the verses her cry was the same as the cry of everybody, 'Give me water, water.' As soon as she got a drink, she seemed to revive, and then she would begin her singing again. The last hymn she sang was 'Nearer, my God, to Thee,' and then she died where she was sitting.

"There was another woman, a Mrs. McAllister. She sat on the deck. She lay still for a while and then called to Clara, the nursemaid. 'Won't you sing a couple of hymns for me?' she said, 'and offer a short prayer, for I am dying.' The nurse knelt right down there in the ashes and began to sing: 'Rock of Ages, cleft for me, Let me hide myself in Thee.' We could hear only snatches of the hymn, for we had work to do, but in each lull we could hear her sweet voice. She sang again: 'Safe in the Arms of Jesus.' Then clasping her hands, she looked up to heaven and offered a short prayer. Mrs. McAllister thanked her and bade her good-bye.

"By this time the air was getting a little purer, so that it was possible to breathe. At the first fierce blast it was so strong and fiery that it struck men dead on the spot. While others were trying to alleviate the suffering of the dying, I went through the different holds of the ship. The hull was tight. What water was in her had come down her hatches when she first heeled over. I sounded under the ship and found twenty-five fathoms of water. Then the second engineer reported that the engines and boiler were safe and that there was no danger of explosion. The second and fourth engineers were seared with fresh scars, but they stayed at their posts to see that the boilers were safe before they left them.

"It was out of the question to get up steam, however, for there was no smokestack and consequently no draught, and if there had been, there was nobody to keep the fires going. Besides, even if the ship had been able to steam off before the wind, the flames in her stern would have swept her decks, instead of burning quietly at the after end of her as she lay at anchor. There was but one thing to do, so, with the assistance of those who could do anything at all, we started to construct a raft.

"First of all we lowered over the side two large skids, full eighteen feet long. Skids are long solid planks bolted together with screw bolts and fitted to the outside of the ship's breast below the davits to prevent chafing when boats are lowered. Then two of us let ourselves down, and after we lashed the skids firmly together, the others passed down lumber from the stock in the cattle pen; these were spiked to the skids and made a raft secure and large enough to carry all who were living on board the ship. We counted the survivors over and found that we had twenty-four living persons. All this took considerable time, and after we got the raft itself constructed, the next thing was to get provisions for her.

"The *Roraima* carried four boats. Three had been destroyed, and the fourth was jammed on the davits. But from here we got oars and oarlocks. We also passed down a compass, lanterns, cases of provisions, a can of oil, and kegs of water. We got everything ready in case the fire from the after end of the ship drove us out before some other means of relief came to us."

Pitifully inadequate though it was, the first rescue operation was under way. From the surrounding *mornes* a group of men had banded together to enter the town. Fernand Clerc refused to join them, arguing that there was still a danger of further catastrophe from Pelée and pointing out that his first consideration, understandably enough, must be for his wife and children. Roger Arnoux, a bachelor with no such claims on his loyalty, led the group on toward St. Pierre.

The zone of destruction covered eight square miles. Within it the destruction of life and habitation had been practically absolute. Where the course of the tornadic blast had traveled across narrow but high-walled valleys, a "haven of refuge" was sometimes found in the lee of one wall, the plane of destruction passing overhead.

On the outskirts of St. Pierre they found hundreds of

bodies, face down, naked for the most part, suggesting that they had in fact been fleeing the town when the blast caught them. Among the bodies they found the barely living Léon Compère-Léandre. Arnoux detailed a couple of men to carry the cobbler to the immediate safety of Mount Parnasse.

By then his group had been joined by a number of other people, including Father Roche and a St. Pierre housewife named Yvette Montferrier. She had been about a mile outside St. Pierre, returning from seeing a relative in Fort-de-France, when the eruption occurred. She had taken shelter in a ditch beside *Le Trace*, and though quite badly scalded, had survived. Now, nearly demented, she was running into the town looking for her husband and children.

To Father Roche she was just one more casualty of the tragedy which had utterly stupefied him. Years later the Jesuit was still to remember his feelings as he went with the others into the ruins:

"It was the smell above all else that shocked. It was a mixture of burning flesh, wood, cinders; a scorched, putrid smell that gripped the throat and numbed and brutalized the brain. And with it came bewilderment and stupefaction. The effect it had on me was primitive and shattering. I looked and did not know what I was seeing. I tried to observe, to remember for later, but I could not. Physicians tell us that when there are too many sound waves, too many light waves, our ears no longer hear, our eyes no longer see. Does something similar perhaps apply to the brain when too many and too brutal impressions strike it at once? I know now the meaning of terror. I know what horror really is."

Even the usually cool and unemotional Roger Arnoux was overcome, as his later testimony reveals: "He who would like to go and share this knowledge of terror, let him go and meditate over the heap of pulverized, formless, putrid things which is all that is left of St. Pierre. Let him go."

To Father Roche, St. Pierre was like "those imitation towns you see in museums, made out of papier-mâché and painted wood. Imagine that an elephant tramples all over one, and it is then set alight, and that finally it is dredged with ash and filth, and you will have what I saw in St. Pierre. Silence enveloped this vista of shattered stone, an unbelievable silence that was broken only by the dull roar of flames. Everywhere there is a porridge of mud and ash, and bodies."

Scorched from the heat, the searchers made their way slowly into the town. The Rue Victor Hugo was no more than a heap of concrete and boulders. Around the Place Bertin rose tier after tier of rubble like some monstrous bulwark. Not a roof remained anywhere. Their eyes followed long lines of half-standing walls, more like the arches of ancient aqueducts than parts of buildings. There was little that rose above a few feet. Bits of polished mosaic paving appeared through the ash, showing where attractive house gardens had been located. All that remained of the Sporting Club were a few of its imposing entrance steps; the rest was a charred stump rising from the general area of smoking debris.

The further into the town they went, the more complete the destruction became. They turned back, their route blocked by corpses, for as Father Roche remembers: "It would have been impossible to have avoided walking over them. They were everywhere, distended by gases, charred. Death seems to have been swift."

From their brief examination of the ruins, Roger Arnoux "concluded that the phenomenon which destroyed it was produced with such suddenness and intensity that there was no chance of escape."

By now Yvette Montferrier had given up any hope of finding her family alive and was in a state of acute shock. Two of the search party had to restrain her from running off into the ruins to seek her husband and children. Even so, she struggled clear near the Jardin des Plantes and rushed into the botanical gardens, pursued by the searchers.

Just inside the ruins of the garden they found a man feebly rolling around in a small pool of water. He was hideously burned. His face was one huge blister, his eyes had been seared from their sockets. Mme. Montferrier was kneeling beside him, shouting had he seen her husband.

Gently, the rescuers pulled her away, and Roger Arnoux eased the man from the puddle.

"I know him," said Father Roche. "It is Professor Landes."

Moments later Gaston Landes died in the gardens which had given him so much pleasure.

It was just after nine o'clock. His death brought the toll in St. Pierre in the last hour to 29,933 men, women, and children. Nearly all of them had died in that moment, at two minutes past eight o'clock, when Mount Pelée had blown an area roughly a hundred yards square out of its side in a violent blast that is still unique in the history of volcanic eruptions.

On board the *Roddam*, Captain Edward Freeman had regained control of his ship. His compass had broken, and he was barely managing to steer, using his elbows to move the wheel. But caught in the current, the *Roddam* was being carried out to sea and southward, back toward St. Lucia, from which it had sailed nine hours before. Below him, on the deck, the dead had been laid out, twenty-six bodies in all.

All day, like a phanton ship—"a ghastly, ghostly apparition"—the *Roddam* drifted southward, until finally at dusk Captain Freeman nosed her into the harbor at Castries on St. Lucia. A customs boat drew alongside, and an officer demanded: "Where have you come from?"

"From the Gates of Hell," said a weary Captain Freeman. He would spend three weeks in a hospital, and on returning to England, he would receive a medal for his heroism.

Inside those "gates of hell," in the roadstead, Ellery Scott had given up his fight to save the *Roraima*. He gave the order to abandon ship: "Gradually we got the passengers over the side to the raft. After the passengers came the crew, then the officers. The second engineer went ahead, and I followed last. Just before I went over the side, I caught sight of a solitary sheep, the last of thirty which had been swept overboard. The poor creature was bleating pitifully. So I went back and laid open its head with an axe, which seemed to me an act of mercy.

"We left behind on board about twenty or thirty dead bodies. My own poor boy was there somewhere. From the moment of the explosion, I never saw him again. He was a likely young fellow, and used to say that some day he would have a ship of his own and would take me along as mate.

"As we looked back, we saw a strange thing. A common reed chair, such as you often see on the deck of a transatlantic liner, was hanging in the air to the ship's stern. It had been fastened to the after flagmast and braced below so that it hung off in space just beyond the reach of the flames. Some poor wretch had rigged it there and had sat in it, clearly afraid to jump on account of the fierce rush of the volcanic currents below. He must have suffered terribly before he dropped from his perch and went overboard as we made to rescue him. He sank beneath the surface and was never seen again."

Far out at sea tow ships steamed purposefully southward, bypassing Martinique. They were part of a mighty rescue task force heading for the island of St. Vincent. Since the Soufrière had erupted the previous day, the great maritime powers of the world, led by Great Britain, America, Japan, and Germany, had offered more than enough help to an island, which, though cruelly hit, had in no way suffered like Martinique.

The first real picture of the total disaster which had hit St. Pierre came from Captain Jules Thirion, when the *Pouyer-Quertier* arrived in Fort-de-France three hours after the eruption.

On the way the cable ship had passed the *Suchet*, at last making "all speed" for St. Pierre. In the wake of the warship was an assorted flotilla of ships, carrying food and first aid equipment, none of which would be of any use.

Waiting for Captain Thirion on the dock was Acting Governor L'Heurre who had set up his operation headquarters in the telegraph office. He had instructed an operator to keep calling St. Pierre, "in the hope that a reply would come." By road he had dispatched troops, doctors, nurses, and food supplies. With this army had gone the Vicar-General.

But for three hours Edouard L'Heurre had waited for any real news. A few minutes before the *Pouyer-Quertier* arrived, an urgent cable, in answer to his distress signal, had come from the Ministry of Colonies in Paris. It ordered:

CABLE URGENTLY DETAILED INFORMATION ON ST. PIERRE CATASTROPHE AND NAMES OF VICTIMS AND SURVIVORS HAVING RELATIVES IN FRANCE.

It would seem that the bulk of the dead, the mulattos, the Negroes, and the Creoles, who had no relatives in France, were not of immediate interest to the French Government.

Equipped with Captain Thirion's lucid description of the fate which had overcome St. Pierre, Acting Governor L'Heurre sent the following reply:

"FIRESTORM DESTROYED ST. PIERRE AND SHIPS IN ROAD ABOUT EIGHT A.M. PRESUME ENTIRE POPULATION ANNIHILATED. NO NEWS OF GOVERNOR MOUTTET OR MME. MOUTTET."

In St. Pierre, the Vicar-General, Gabriel Parel, Dr. Emile L'Herminer, and Philip Rose, a Government chemist, were all convinced that the Governor, his wife, and the ill-fated Commission of Inquiry had all been incinerated in the Hôtel de l'Indépendance.

With a column of marines as their escort the trio had arrived in St. Pierre at midday. By then most of the fires had died down. Pelée, its summit several hundred feet lower now, was sending only harmless looking vapor whiffs into the clear sky.

In the town itself scores of people were searching for survivors and getting in the way of soldiers and others more skilled in rescue work.

Nearly fifty soldiers had been directed to dig over the ruins of the Hôtel de l'Indépendance. A number of bodies were recovered. None of them were identified as Governor Mouttet or his wife, nor any member of the Governor's commission. As each body was brought out, Dr. L'Herminer noted their condition.

To the official inquiry later he reported: "They were burnt out. The cause of death was asphyxia. Without a detailed examination of the skull formation, it was impossible to tell whether they had been white or Negro. The action of the fire was excessively curious. While some parts of the body were completely destroyed, other regions were merely blackened. Thus, for example, the sexes were respected. There was erection in the case of some of the male corpses. But not all. It was rather the exception. Many women, those who one would guess were young, had the breasts intact. All the corpses were naked. Scalped, depilatated. In many the abdomen had burst and the intestines had gushed out, unburned. They were of a purplish red color, the color of wine. The different degrees of burning may be explained by the action of the explosion on the muscles. Under this action the strongest bodies spontaneously contracted. The limbs were under tension. The weakest were forced into extension, and the most exposed were burned worse than the rest. This explains the attitude of the overwhelming majority of the bodies. Limbs flexed, chest distended, head thrown back, neck out."

But not one of those bodies was that of Louis Mouttet. In spite of a week of continuous searching, his corpse was never found.

Fittingly enough the epitaph of St. Pierre may be found in the Vicar-General's diary:

"With what profound emotion I raise my hand above these thirty thousand souls so suddenly mowed down, buried in this terrible tomb to sleep the sleep of eternity. Beloved and unfortunate victims! Priests, old men and women, sisters of charity, children, young girls, fallen so tragically, we weep for you, we the unhappy survivors of this desolation. While you, purified by the particular virtue and the exceptional merits of this horrible sacrifice, have

risen on this day of the triumph of your God to triump[h]
with Him and to receive from His own hand the crown [of]
glory. It is in this hope that we seek the strength to surviv[e]
you."

Epilogue

No one was officially blamed or held responsible, even partially responsible, for the deaths of 29,933 people in St. Pierre on that Thursday morning of May 8, 1902.

Pierre Louis Albert Decrais, the Minister for Colonies, Louis Marius Mouttet, Andréus Hurard, Professor Gaston Théodore Landes, Senator Amédee Knight, Roger Fouché, Jules Gerbault, Alfred Percin, Osman Duquesnay—all of whom had in some measure had a hand in a tragedy which could have been averted—were all to escape official censure. They were not so lucky with the press of the day. A number of writers scourged not only the guilty who had died, but those who had survived. But not one of the survivors took any legal action against his critics.

The principal survivors were to meet various fates:

Fernand Clerc, his wife, and children moved to a new plantation at Vive, in the center of the island. He was a member of the Chamber of Deputies in 1920, and died in 1921.

Senator Amédee Knight returned to Paris, where he continued to represent the island for another term.

Curé Mary, the priest of Morne Rouge, died in a hospital in Fort-de-France on September 1, 1902 as a result of burns he had received two days earlier when Pelée erupted again, destroying most of the village of Morne Rouge.

Father Alte Roche remained on the island, becoming a priest in Fort-de-France.

The Vicar-General, Gabriel Parel, remained on the island until his death the next year.

The de Jaunville estate was completely destroyed, but Colette de Jaunville survived long enough to be taken to the hospital in Fort-de-France, where she died on the evening of May 8th. At the time, René Cottrell was in St. Pierre searching for her.

Léon Compère-Léandre died in 1936, also a bachelor.

Auguste Ciparis, as mentioned earlier, became an exhibit in the Barnum and Bailey Circus. He died in 1929.

Suzette Lavenière was to administer her estate until her death in 1939 at the age of sixty-one.

Ellery Scott, the chief officer of the *Roraima,* retired from the sea in 1923.

Captain Edward Freeman, master of the *Roddam,* traveled home to England on the same steamer that took the nuns from Morne Rouge to France. He retired from the sea in 1928.

Boverat, the servant of James Japp, the British Consul, took up a new post in service with a private family in Fort-de-France. He died in 1919.

Eleven days after the eruption, the bodies of James Japp and Thomas Prentiss were finally found in their devastated Residencies by British and American naval rescue teams. They were placed in coffins, and on May 19, Thomas Prentiss's remains were transported by an American warship to Fort-de-France, where he was buried with full military honors.

James Japp was not so well looked after. Joseph Durival, aged 72 and curator of the Volcanological Museum in St. Pierre, told us that "as M. Japp's body was being carried on the shoulders of the British officers out of the town, Pelée threatened again. Immediately, the seamen dropped the coffin and fled to their ship, leaving Japp's remains to the mercy of Pelée."

One should not be too critical of the sailors' actions. The next day, May 20, Pelée exploded more fiercely than

Epilogue

ever. Fortunately, no living person was in the town as the volcano flattened anything that had withstood the May 8 eruption, and buried forever British Consul James Japp—"without mercy," and without military honors.

Even before Japp's body was identified on May 19, British Vice-Consul Devaux, based on Guadaloupe, applied to the British Foreign Secretary, the Marquis of Lansdowne, to be allowed to replace Japp. Whether it was the haste of his application or the florid manner in which it was penned, the Marquis wrote a curt refusal.

Aid, when it came, was world-wide. President Theodore Roosevelt, on behalf of the American Government, gave $250,000. Other donations from sovereigns and Heads of State included: Emperor of Russia $25,000; King of England $2,500; Emperor of Austria $2,500; King of Italy $2,500; Princess Waldemar of Denmark $2,033. The President of the Republic of France gave $2,000, less than the Sultan of Turkey, who gave $3,000, and slightly more than the King of Siam, who gave $1,500. The lowest donation came from President Kruger of South Africa, who gave about $80.

By November, 1902, the total subscribed had risen to $879,162, over half of which had been donated by the French Government ($500,000). Of this figure $231,370 was distributed to help refugees, for the whole island had been affected; crops had withered, and even in the southern end of the island houses were coated with ash. Much was spent on rebuilding St. Pierre itself. The remainder, $647,793, was invested in French Treasury Bonds.

Among the biggest memorial services were those held in Paris, on May 18, 1902; London, on May 25, 1902; and Rome, on June 7, 1902.

St. Pierre has been partially rebuilt beside the remains of the 1902 disaster. But with a population of only seven thousand, the town today is of secondary importance to Fort-de-France.

Since 1902 Pelée has erupted from time to time, with serious explosions in 1929 and 1930, when St. Pierre was again threatened. M. Raphaël Petit, Prefect of the Haute-Loire, and a former Prefect of Martinique, recalls, as a child, seeing the people flee: "a spectacle I was to relive again in a different setting but with the same miserable and illogical side on the roads of Belgium and France in 1940. . . ."

Tourist brochures say that Mount Pelée is now extinct. This may be, but she is certainly not harmless. One of the most poignant stories of the volcano's continuing menace was told to us by Madame Marie Dufrénois, now ninety-two years old.

Madame Dufrénois lives in Morne Rouge, which has been rebuilt since its destruction in August, 1902. She herself has a vivid recollection of that fateful week in May, 1902, and after describing these events to us, she went on to tell us of the tragedy which befell her son at the age of thirty-three. In 1937 he had climbed Pelée to look down into its crater. He slipped and fell 250 feet to its bottom. There he remained paralyzed through the night. The next day he was discovered. A doctor was lowered to him, and eventually he was lifted out of the crater and taken to a hospital, where he died thirteen hours later on the operating table.

Perhaps Pelée will always claim victims.

Bibliography

Bullard, Fred Mason	*Volcanoes*	Nelson London 1962
Cole, Hubert	*Josephine*	Wm. Heinemann Ltd. London 1962
Cartland, Barbara	*Josephine—Empress of France*	Hutchinson & Co. Ltd. London 1961
Chapman, Guy	*The Dreyfus Case*	Hart Davis London 1955
Creole, Coeur	*Saint Pierre, Martinique 1635 1902.*	Paris 1905
Croze, F. de	*Le Désastre de la Martinique*	Barbou, Limoges 1903
Deerr, Noel	*Cane Sugar*	Norman Rodger Manchester 1910
Deschamps, Phillipe	*Les Cataclysmes de la Martinique*	Passage Choiseul Paris 1903
Duchateau, Roger E.	*Une Histoire Vécue des Cataclysmes de la Martinique*	Desclée Lille 1907
Fermor, Patrick L.	*The Violins of Saint-Jacques*	John Murray Ltd. London 1953
Garnham, S. A. and Hadfield, Robert	*The Submarine Cable*	Sampson Low, Marston & Co. Ltd. London
Frierson, Francis	*Famous French Crimes.*	Muller, London 1959
Heilprin, Angelo	*The Eruption of Pelée*	J. B. Lippincott Co., Philadelphia 1908

	Mont Pelée and The Tragedy of Martinique	1903
Hearn, Lafcadio	*Two Years in the French West Indies*	Harper & Brothers London 1890
Hess, Jean	*La Catastrophe de la Martinique*	Charpentier et Fasquelle Paris 1902
Hand, T. W.	*The Last Days of St. Pierre*	Ottawa 1902
Kennan, George	*St. Pierre and Mt. Pelée Through the Stereoscope*	The Outlook Co. New York 1903
	The Tragedy of Pelée	The Outlook Co. New York 1902
Lane, Frank W.	*The Elements Rage*	1966
Lacroix, A.	*La Montagne Pelée Après Ses Éruptions*	Académie des Sciences Paris 1908
	La Montagne Pelée et Ses Éruptions	Académie des Sciences Paris 1904
	Sur l'Éruption de la Martinique	Comptes Rendus Vol. CXXXV 1902
Lascroux	*La Martinique Avant et Après le Désastre*	New York 1902
Royce, Frederick	*The Burning of St. Pierre and The Eruption of Mt. Pelée*	Continental Publishing Co. New York 1902.
Sterns-Fadelle, F.	*In the Ruins of St. Pierre*	Roseau 1902
Smith, Nicol	*Black Martinique— Red Guiana*	Jarrolds Ltd. London 1943
Welch, J. H. and Taylor, H. E.	*The Destruction of St. Pierre*	New York 1902

Magazines, Periodicals, Papers

Annaures Du Ministre	"Des Colonies"	Paris 1902
American Heritage	"Prelude to Doomsday" (Lately Thomas)	Aug. 1961
Century Magazine	"A Study of Pelée" (R. T. Hill) Vol. LXIV	Sept. 1902
Century Magazine	"Phases of the West Indian Eruption" (Israel C. Russell) Vol. LXIV	Sept. 1902

Bibliography

Century Magazine	"Life In The Doomed City The Last Days of St. Pierre" (Parel)	May—Oct. 1902
Collier's Weekly	"Mont Pelée's Secret"	June 1902
Collier's Illustrated Weekly	"A Visit To The Dead City" (L. Siebold)	June 14, 1902
Cosmopolitan Magazine	"The Eruption of Mont Pelée" (E. S. Scott)	July 1902
The Evening News	"But One Man Lived"	April 21 London 1955
Foreign Office Records	"F. O. 27—3596"	London 1902
Frank Leslie's Popular Monthly.	"The Destruction of The Roraima"	May—Oct. 1902
	"Martinique Supplement"	July 1902
Gazette de la Martinique		March Fort-de-France 1964
		May Fort-de-France 1964
Harper's Weekly	"The Destruction of St. Pierre" (J. S. Flett)	June 7, 1902
His Majesty's Accounts and Papers Vol. LXVI "Correspondence Relating to the Volcanic Eruptions"		Sept. London 1902
Holiday Magazine	"The West Indies" (Peter Fleming)	March 1949
La Revue Francaise No. 193		Paris Nov. 1966
Leslie's Weekly	"Most Fatal and Frightful Disaster of Our Times" (S. S. McKee)	June 12, 1902 June 5, 1902
Lloyds List		April–June London 1902
Lloyds Weekly Shipping Index. Vol 1		London 1902
Lobley, J. Logan	"Volcanic Action and The West Indies Eruptions of 1902" (Paper read before the Victoria Institute, London, April 20, 1902)	
McClure's Magazine	"Mount Pelée in it's Might" (Angelo Heilprin)	Aug. 1902

McClure's Magazine	"Pelée, The Destroyer" (August F. Jaccaci)	Sept. 1902
Martinique and St. Vincent. A Preliminary Report	*Bulletin of the American Museum of Natural History* Vol. XVI	1902
The National Geographic	"Volcanic Eruptions on Martinique and St. Vincent" (Israel C. Russell) Vol. XIII	1902
The National Geographic	"The Shattered Obelisk of Mont Pelée." (Angelo Heilprin) Vol. XVII	1906

(See also *The National Geographic* for June, July, and December of 1902; November of 1903; and August of 1906)

Natural History	"The City of The Dead" (A. L. Koster) Vol. 65	1956
Nature	"The Recent Volcanic Eruptions in The West Indies" Vol. LXVI	London 1902
Pearson's Magazine	"The Awful Doom of St. Pierre" (E. W. Freeman)	London Sept. 1902
Parallèles Supplement	(Mme. Anca Bertrand)	Fort-de-France May 8, 1967
Parallèles No. 4	(Mme. Anca Bertrand)	Fort-de-France 1965
Parallèles No. 7	(Mme. Anca Bertrand)	Fort-de-France 1965
Parallèles No. 14	(Mme. Anca Bertrand)	Fort-de France 1966
Philosophical Transactions of the Royal Society	"Report on the Eruptions of the Soufrière in St. Vincent in 1902, and on a Visit to Montagne Pelée in Martinique" (T. Anderson and J. S. Flett) Series A. Vol. 200	
Readers Digest		Nov. 1961
Strand Magazine	"The Tragedy of Martinique" (Ellery S. Scott)	1902
Scribner's Magazine	"The Martinique Pompeii" (J. R. Church) Vol. XXXII	Vol. XXXII 1902

Youth's Companion	"Mont Pelée and The Ruins of St. Pierre" Vol. LXXVI	Vol. LXXVI 1902

Author's Note

A NUMBER of accounts of May 8, 1902, have been written, mostly soon after the event when it was not possible for the authors to benefit from the many sources of evidence we have been able to call upon.

Inevitably, we have been confronted from time to time with conflicting evidence, as in the Guérin disaster, when some put the total dead as high as 400; as in the over-all death roll, which has been reputed to be 50,000; or in the number of survivors in the town, which has fluctuated between none and four.

Where we had to make a choice of fact, we relied most heavily on the official reports of the Académie des Sciences, the Royal Society, and the American Museum of Natural History.

Whenever there was a choice of interpretation, we relied on our own judgment. In these isolated instances, the decision was ours alone, and none of those who so generously helped us during our research should be thought to have automatically endorsed our portrayal of the events as we have described them.

G. T. & M. M. W.
London 1968

Background to the Book

THE SEED FOR THIS BOOK was sown in October 1965 when, on Channel-One, B.B.C. Television transmitted a major film documentary entitled "The Days The World Went Mad." This program written by Gordon Thomas and produced and directed by Max Morgan Witts, received wide critical acclaim.

"The Days The World Went Mad" described ten natural disasters, and included a short account of the volcanic eruption of May 8, 1902 in Martinique. Since then, the authors have continued their research into this particular disaster. Between them, they have traveled thousands of miles, collated material from all over the world, and interviewed, personally or by letter, hundreds of persons.

ABOUT THE AUTHORS

GORDON THOMAS and MAX MORGAN WITTS are unrivaled in their ability to get the facts and bring dramatic moments in history to life. For each of their successful books, they have travelled thousands of miles and interviewed hundreds of people, tracking down eyewitness accounts and unearthing information. The record speaks for itself. Their novel, *Voyage of the Damned* was made into a major motion picture and was a Literary Guild choice; *The San Francisco Earthquake* was a Book-of-the-Month Club Main Selection; *Guernica* was a Literary Guild Alternate Selection and now, *When Time Ran Out* is also a major motion picture.

RELAX!
SIT DOWN
and Catch Up On Your Reading!

☐	12982	**EXORCIST** by William Peter Blatty	$2.50
☐	12626	**THE WOLFEN** by Whitley Strieber	$2.50
☐	13098	**THE MATARESE CIRCLE** by Robert Ludlum	$3.50
☐	12206	**THE HOLCROFT COVENANT** by Robert Ludlum	$2.75
☐	13688	**TRINITY** by Leon Uris	$3.50
☐	13899	**THE MEDITERRANEAN CAPER** by Clive Cussler	$2.75
☐	12152	**DAYS OF WINTER** by Cynthia Freeman	$2.50
☐	13176	**WHEELS** by Arthur Hailey	$2.75
☐	13028	**OVERLOAD** by Arthur Hailey	$2.95
☐	13220	**A MURDER OF QUALITY** by John Le Carre	$2.25
☐	11745	**THE HONOURABLE SCHOOLBOY** by John Le Carre	$2.75
☐	12313	**FIELDS OF FIRE** by James Webb	$2.50
☐	13880	**RAISE THE TITANIC!** by Clive Cussler	$2.75
☐	12855	**YARGO** by Jacqueline Susann	$2.50
☐	13186	**THE LOVE MACHINE** by Jacqueline Susann	$2.50
☐	12941	**DRAGONARD** by Rupert Gilchrist	$2.25
☐	13284	**ICEBERG** by Clive Cussler	$2.50
☐	12810	**VIXEN 03** by Clive Cussler	$2.75
☐	12151	**ICE!** by Arnold Federbush	$2.25
☐	11820	**FIREFOX** by Craig Thomas	$2.50
☐	12691	**WOLFSBANE** by Craig Thomas	$2.50
☐	13017	**THE CHINA SYNDROME** by Burton Wohl	$1.95
☐	12989	**THE ODESSA FILE** by Frederick Forsyth	$2.50

Buy them at your local bookstore or use this handy coupon for ordering:

Bantam Books, Inc., Dept. FBB, 414 East Golf Road, Des Plaines, Ill. 60016

Please send me the books I have checked above. I am enclosing $_____ (please add $1.00 to cover postage and handling). Send check or money order —no cash or C.O.D.'s please.

Mr/Mrs/Miss _____

Address _____

City _____ State/Zip _____

FBB—4/80
Please allow four to six weeks for delivery. This offer expires 10/80.

Bantam Book Catalog

Here's your up-to-the-minute listing of over 1,400 titles by your favorite authors.

This illustrated, large format catalog gives a description of each title. For your convenience, it is divided into categories in fiction and non-fiction—gothics, science fiction, westerns, mysteries, cookbooks, mysticism and occult, biographies, history, family living, health, psychology, art.

So don't delay—take advantage of this special opportunity to increase your reading pleasure.

Just send us your name and address and 50¢ (to help defray postage and handling costs).

BANTAM BOOKS, INC.
Dept. FC, 414 East Golf Road, Des Plaines, Ill. 60016

Mr./Mrs./Miss_____
(please print)
Address_____
City_____ State_____ Zip_____

Do you know someone who enjoys books? Just give us their names and addresses and we'll send them a catalog too!

Mr./Mrs./Miss_____
Address_____
City_____ State_____ Zip_____

Mr./Mrs./Miss_____
Address_____
City_____ State_____ Zip_____

FC—9/78